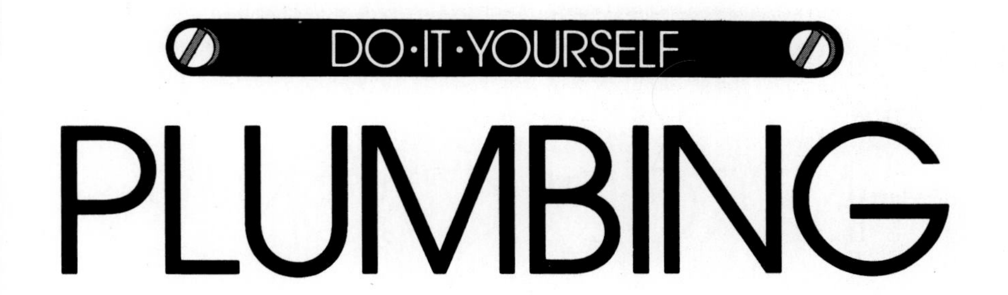

Excerpted from the Better Homes and Gardens®
Complete Guide to Home Repair, Maintenance and Improvement

First Edition. First Printing.
ISBN: 0-696-01866-7

INTRODUCTION

For many do-it-yourselfers, plumbing is the last frontier. But most plumbing repairs are well within the capabilities of novices, and with today's materials, even installation of a new fixture lends itself to do-it-yourself execution. In Better Homes and Gardens® *Do-It-Yourself Plumbing*, you'll learn how to solve the problems you face most often, from drippy faucets to clanging pipes. We'll show you how to choose materials for plumbing extensions, how to rough in new lines, and how to install fixtures. With hundreds of concise illustrations and step-by-step instructions, *Do-It-Yourself Plumbing* gives you the information—and the confidence—you need to build your plumbing skills.

Better Homes and Gardens® Books
Editor: Gerald M. Knox
Art Director: Ernest Shelton
Managing Editor: David A. Kirchner
Editorial Project Managers: Liz Anderson, James D. Blume, Marsha Jahns, Jennifer Speer Ramundt, Angela Renkoski

Do-It-Yourself Plumbing
Editors: Gayle Goodson Butler and James A. Hufnagel
Editorial Project Manager: Liz Anderson
Electronic Text Processor: Paula Forest

Complete Guide to Home Repair, Maintenance and Improvement
Project Editors: Larry Clayton, Noel Seney
Contributing Project Editor: James A. Hufnagel
Graphic Designer: Richard Lewis
Illustrations: Graphic Center

CONTENTS

PLUMBING

The oldest and simplest of a home's systems, plumbing seems mysterious only until you realize it relies on just two physical principles—pressure and gravity. Turn on a faucet full blast and you can feel pressure pushing water through pipes to your fixtures; pull a drain plug and gravity carries the water away.

Because of this simplicity, the hidden parts of your plumbing system—its pipes and the fittings that tie them together—rarely give you trouble. When something does go wrong, it usually happens at a fixture or in a drain pipe, either of which you can easily service yourself.

This book begins by introducing you to your system's inner workings, tells you how to cope with plumbing emergencies and repairs, then goes on to illustrate what you need to know about upgrading your home's waterways.

GETTING TO KNOW YOUR SYSTEM

Your home's water supply, which comes either from the city or a well, enters via a sizable pipe. If you're on city water, this pipe connects to a *meter* that reads the amount of water entering. Next to the meter is a *shutoff valve* that lets you stop all flow of water, if necessary. The supply pipe then travels to a *water heater,* or in the case of private systems, to a *pressure tank* and then to a water heater.

From the heater emanates a *hot water supply line.* This line and a *cold water supply* run parallel throughout your home to serve the various *fixtures* (water faucets, lavatories, bathtubs, toilets, etc.) and *appliances,* such as clothes- and dishwashers. The supply lines are under constant pressure—usually 50 to 60 pounds per square inch.

Another system of pipes, the *drain-waste-vent pipes* (DWV), carries away water and waste to a *city sewer* or a *septic system,* and vents potentially harmful gases to the outside. Not under pressure, these pipes depend on gravity to perform their function.

ANATOMY OF A PLUMBING SYSTEM

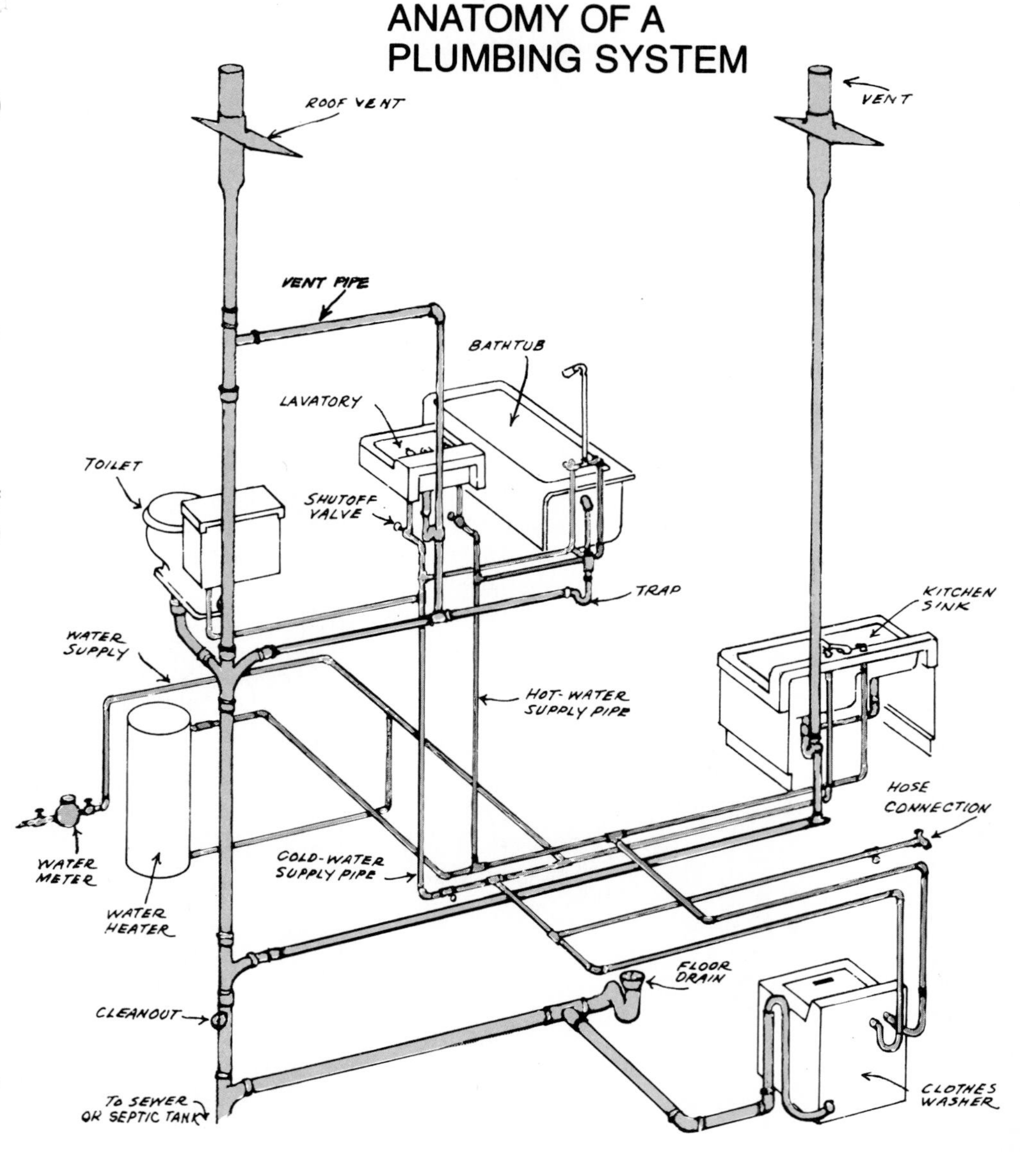

SHUTOFF VALVES

The cold and hot water supply system in your home features a series of *shutoff valves*—sometimes called *stops*. Think of them as on/off switches that provide you with an easy-to-find, quick-to-close turnoff network in the event the piping system springs a leak or if you want to make repairs or replace any of the system's components.

Look for stops near the point where supply lines enter a fixture or an appliance. If you don't find any there, use the main shutoff on the meter's street side. Closing this valve turns off your entire water system.

Water supply lines at sinks, lavatories, and flush tanks also may be equipped with shutoff valves. Water heaters have one, too.

With some fixtures, such as tubs, the shutoff may be beneath the floor, as shown, or hidden behind an access panel.

If your home has a meter, you'll find two main shutoffs. Use the one on the supply side so pressure doesn't damage the meter.

TRAPS

Traps perform the useful function of preventing dangerous gases generated in the drain-waste lines from backing up and infiltrating your home. These simple, yet effective devices do this by creating an automatic water seal (see sketches at right), which forces the gases to rise up the soil stack. Running water flushes the trap, but gravity ensures that some will always remain.

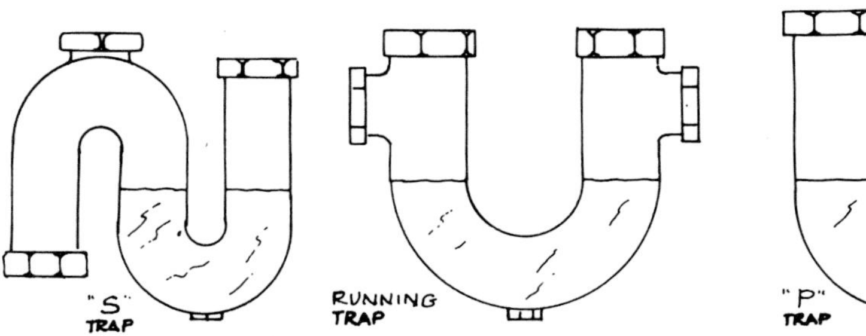

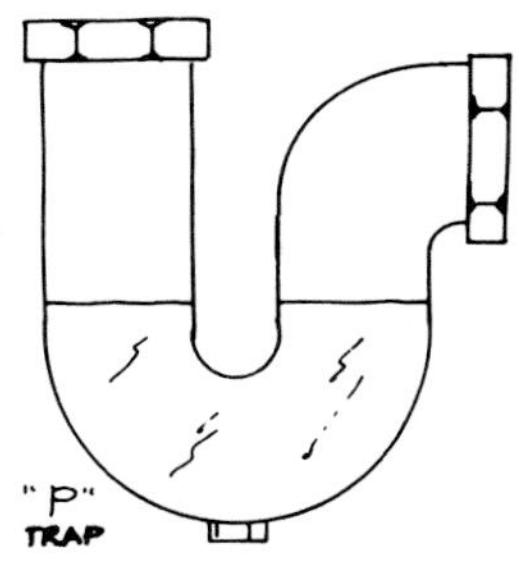

PLUMBING CODES

For everyone's protection, a national plumbing code spells out specific guidelines for all plumbing operations. Most cities, towns, and communities have adopted this code and amended it to fulfill local requirements and conditions.

Before you install new plumbing, read or ask about the plumbing codes in your area to make sure your intended project conforms with the code. A permit may be required for new plumbing work. You'll probably also have to have a plumbing inspector check your installation.

READING A METER

Whether your meter is a direct-reading type like the one shown at near right, or has a series of dials like the one at far right, determining consumption is an exercise in subtraction.

With dial types, note the position of each pointer, wait a few days, then note their positions again. Subtract the first reading from the second for the number of gallons of water used.

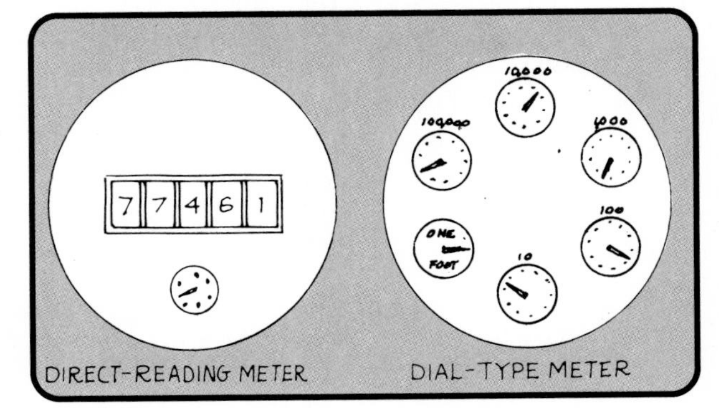

If you suspect a water leak but can't locate it, give your water meter a leak test. First, turn off all the faucets and water-using appliances in your home. Then take a look at the scales on your water meter (see sketch at left). Watch the one-cubic-foot scale for 20 minutes or so. If the dial moves at all during this time, the water supply system is leaking, probably behind a wall or underground.

SOLVING PLUMBING PROBLEMS

When confronted with a plumbing problem—even one of those "nuisance" repairs such as fixing a dripping faucet or unclogging a drain—too many people simply throw up their hands and call a plumber. Then, often as not, they wait hours or even days for a repair that takes only a few minutes, but costs a lot of money.

If that's happened at your house, the next 20 pages are for you. They delve into just about any difficulty you're likely to encounter, explain the relatively simple components you'll be dealing with, and present the know-how you'll need to handle a situation confidently and effectively. And you don't need a whole toolbox full of gear to get started—just the basics shown below.

After you've mastered the fundamentals of plumbing repair, you may then want to go on and try some of the improvements covered later.

BASIC TOOLS

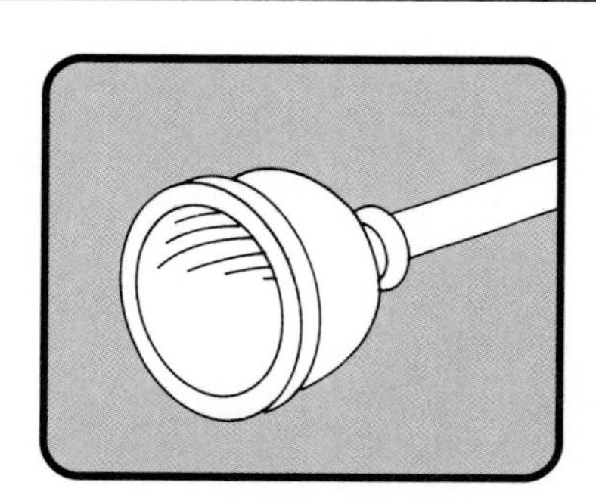

A "plumber's helper"—also called a *plunger* or *force cup*—dislodges debris from drains by creating suction.

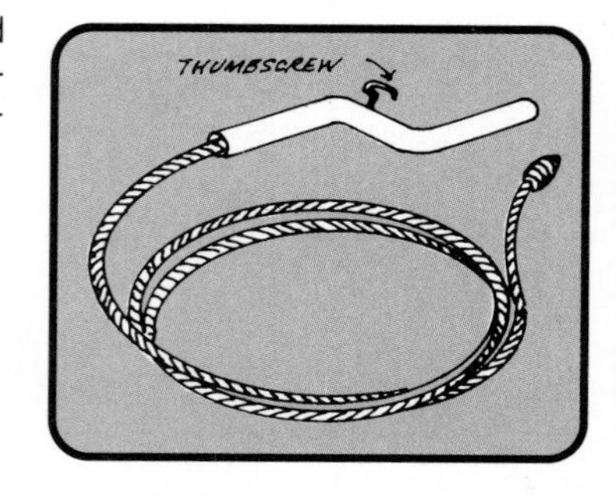

A flexible *plumber's snake* can be threaded through drainpipes. Locking the thumbscrew lets you crank to break up debris.

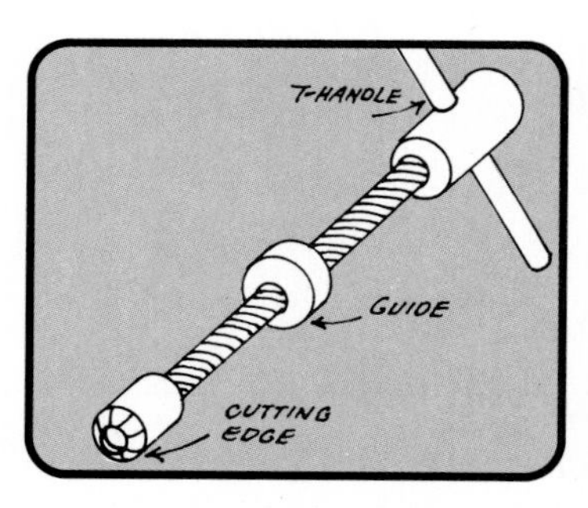

To clean corroded valve seats in faucets, you'll need a valve-seating tool (seat grinder). The cutting end fits into the faucet.

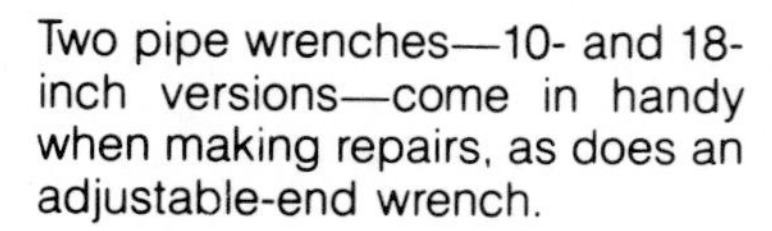

Two pipe wrenches—10- and 18-inch versions—come in handy when making repairs, as does an adjustable-end wrench.

Chemical drain cleaners, used periodically, keep drains open and functioning perfectly. They're tops for preventive maintenance.

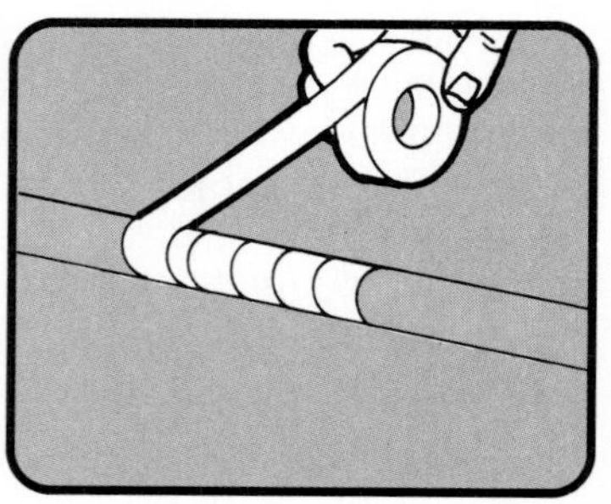

Plastic electrician's tape will temporarily stop a pinhole leak in a water supply pipe. The bad pipe must be replaced, however.

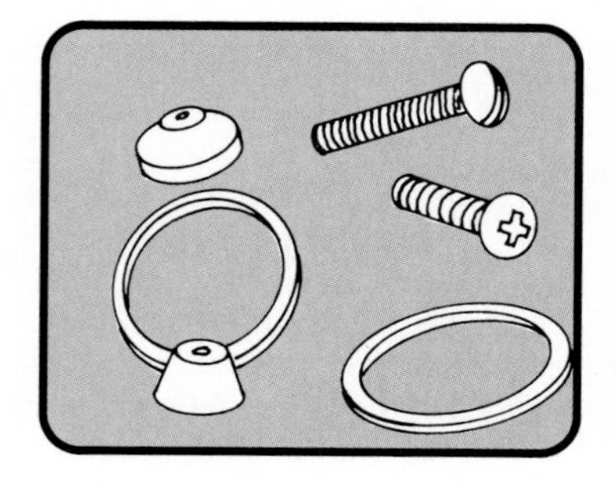

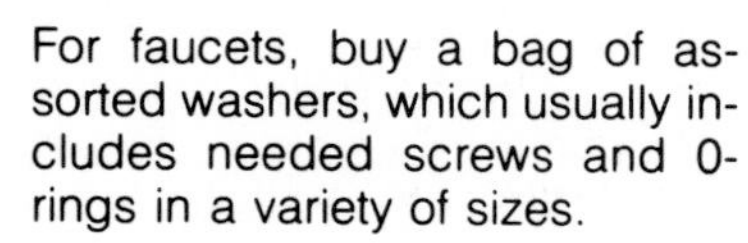

For faucets, buy a bag of assorted washers, which usually includes needed screws and O-rings in a variety of sizes.

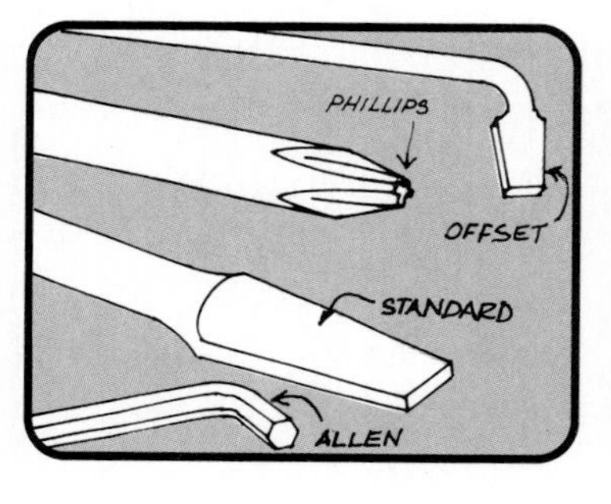

For plumbing fixtures, you'll need screwdrivers with standard, Phillips, and *offset*-type blades. *Allen* wrenches are also handy.

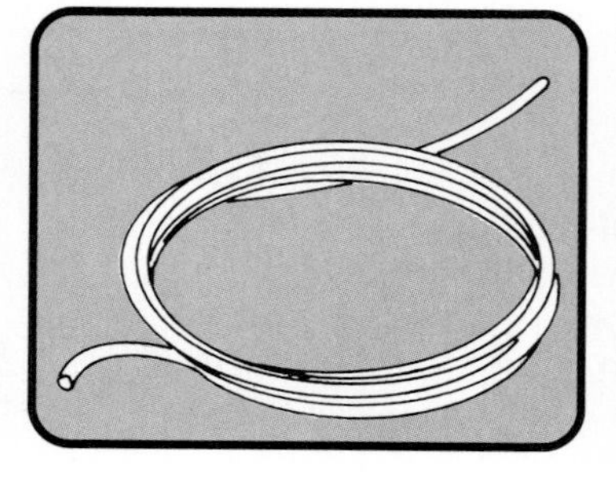

Packing for faucet nuts looks like heavy twine that's been coated with black or brown caulk. Keep the packing sealed.

Use joint compound (pipe dope) or pipe tape to seal threads when you reassemble an old connection or assemble new piping.

FIRST AID FOR LEAKS

Sometimes it almost seems as if pipes schedule leaks to correspond with the closing hours of hardware stores and plumbing outlets just to confound you. If you find yourself in this untimely predicament (and you will sometime or other), and if the flow isn't of gusher magnitude, you can get by temporarily by making an emergency patch.

The patch can be any material that will stop the flow of water: a piece of rubber and a C-clamp, several layers of plastic tape, even a length of garden hose split and tied around the pipe. Far better, though, is an emergency patch kit. Several inexpensive solutions are shown below.

When you notice a leak, turn off the water at the main supply valve or a shutoff valve (see page 5) first thing. This takes pressure off the line. After this, diagnose the exact damage and administer the appropriate repair.

If the leak is more of a drip than a squirt, it may be water condensation, not a break in the pipe, joint, or fitting. If this is the problem, dry up the pipes and fittings with insulating sleeves that you cut to length, wrap around a pipe, and seal or tape in place.

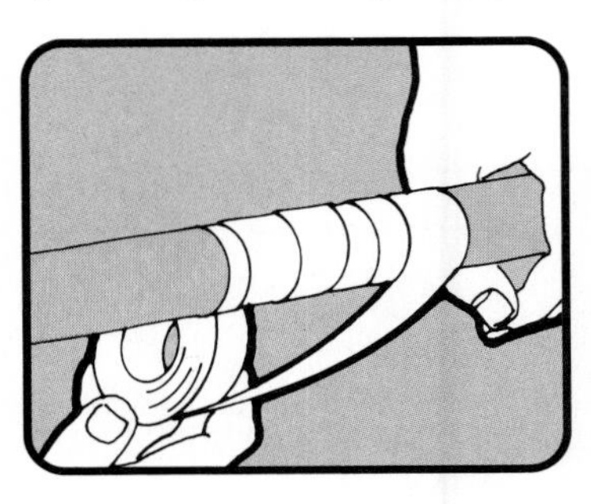

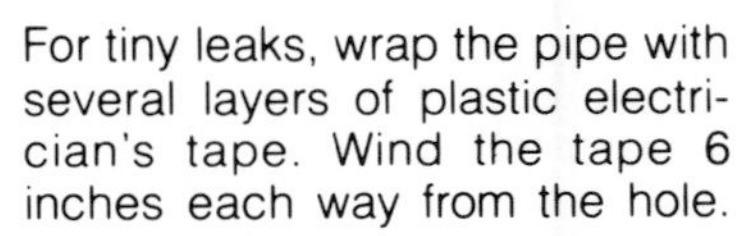

For tiny leaks, wrap the pipe with several layers of plastic electrician's tape. Wind the tape 6 inches each way from the hole.

Epoxy putty works well for leaks at connections. Spread it around the leak with a putty knife; it dries quickly.

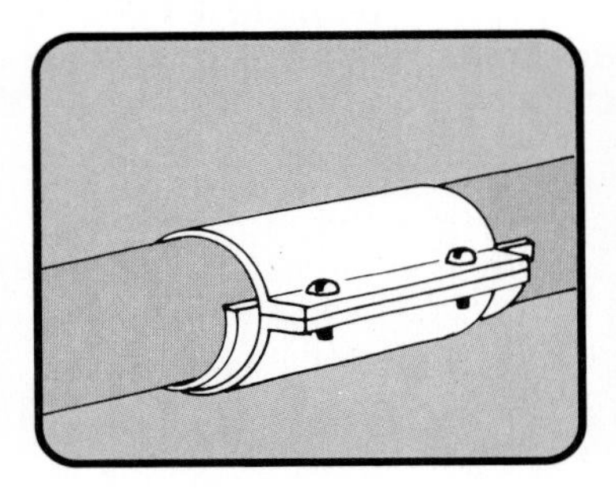

This special metal clamp has a rubberlike inner lining insert. You can tighten the clamp with a screwdriver.

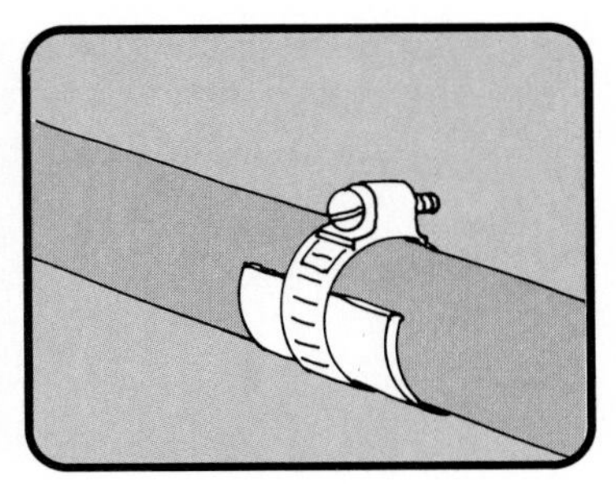

An auto hose clamp and a piece of rubber make an excellent emergency leak stopper until you can replace the leaky pipe.

THAWING FROZEN PIPES

Having water pipes that freeze during the winter not only is frustrating, it's downright dangerous. If you don't thaw them soon after they freeze, you're asking for burst water lines, and sooner or later your request will be granted.

Before attempting any of the thawing techniques shown here, be sure to open the faucet the frozen pipe supplies. The steam created by the heat you'll apply must be able to escape. Otherwise, your pipe will burst.

If the frozen pipe is behind a wall or in a ceiling or floor, you're best off placing a heat lamp as near the faucet as possible.

If you use a propane torch to thaw a frozen pipe, be extremely careful of fire. Don't use open flame to thaw frozen pipes behind walls, ceilings, or floors, or near gas lines.

If freezing pipes are a constant threat in your area, wrap the pipes with insulation made in narrow widths especially for pipes, or buy wool felt, plastic foam, air-cell asbestos, or pipe jackets to minimize freezing.

Sometimes other tactics work, too. If pipes under a kitchen sink that's on an outside wall are a problem, try putting a small lamp down there—or just open the cabinet doors and let room heat warm the pipes.

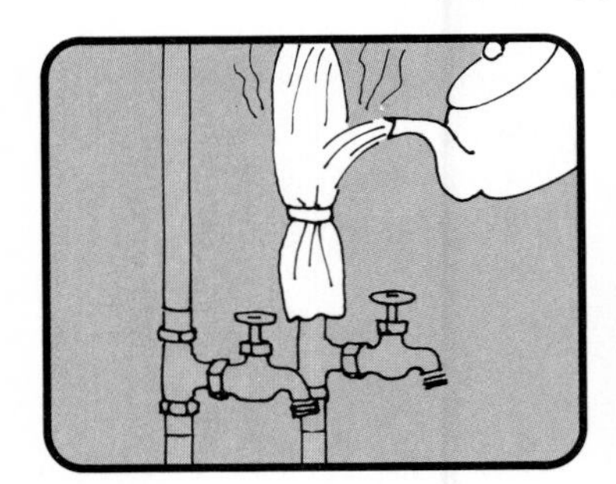

Although messy, you can thaw a pipe by wrapping it with layers of cloth, tying the cloth to the pipe, and pouring hot water on it.

A heat lamp is excellent for exposed or concealed pipes. But protect other materials around the pipe—heat lamps can scorch.

For exposed pipes, use a propane torch, but watch out for fire. Work from the faucet toward the frozen area. Open the faucet.

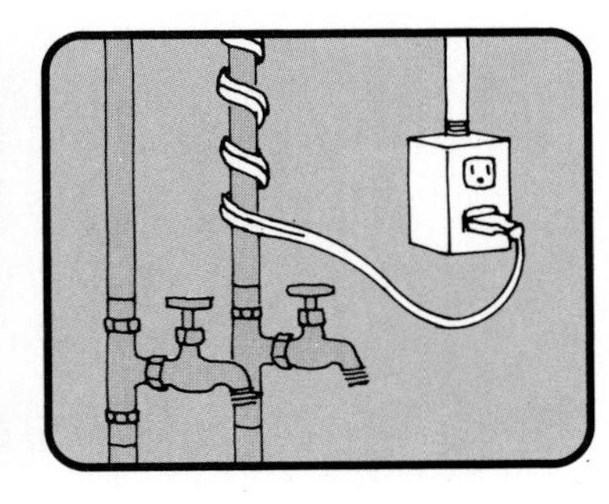

Electric heat tape runs on house current. Wrap it around the pipe and plug it into an outlet. A thermostat controls the heat.

OPENING CLOGGED DRAINS

Impossibly stopped-up drains call for a professional with the electric equipment only pros can afford. But fortunately, most drain problems don't fall into this category, and often you can handle them yourself.

First realize that your home has three types of drains: fixture drains such as those at sinks and toilets; main drains, which lead from the fixture drains to the main pipe that carries waste from your home; and sewer drains, which run underground to the community sewer or septic tank.

Your problem can originate in any of the three, so your most immediate chore is to locate the blockage. Almost always it will be in or next to a pipe connection that makes a turn, or in a trap.

To pinpoint the difficulty, open a faucet at each sink, tub, or other fixture—but don't flush a toilet; it could overflow. If only one fixture is stopped up, the problem is right there or nearby. If two or more fixtures won't clear, something has lodged itself in a main drain. And if no drains work, the blockage is farther down the line—either near the point where the main drain or drains connect to the sewer drain, or in the sewer drain itself.

Bearing in mind that waste water always flows downward through pipes of increasingly larger diameter lets you logically ferret out an obstruction you'll almost certainly never see. The drawings here and on the opposite page show what to do once you've found it.

Sinks

Hair, bits of soap, and other debris can gum up a lavatory stopper. To remove some types just turn and lift.

A plunger with a molded suction cup is ideal for toilets or rounded lavatory bowls. Flat plungers work best on flat surfaces.

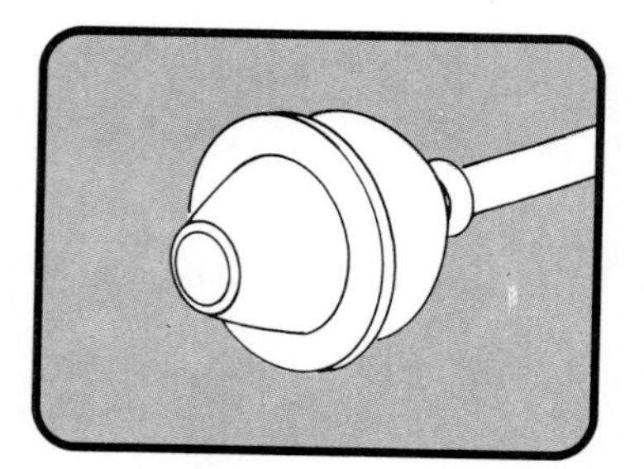

If a sink has an overflow outlet, plug it with a cloth and make sure the plunger seals tightly over the drain outlet.

If a plunger won't work, try an auger snake. Thread it down and through the trap, or open the cleanout and work from there.

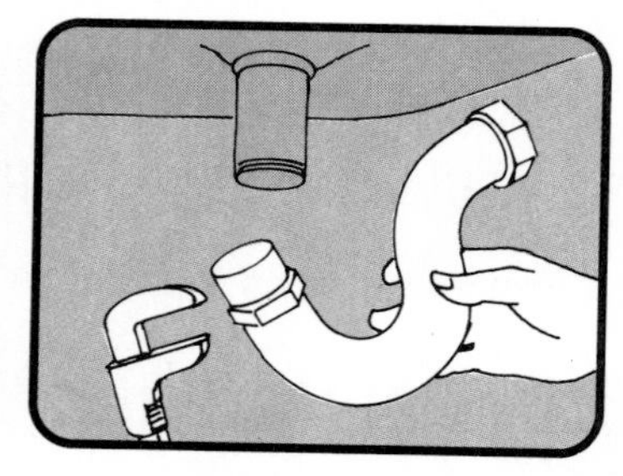

If augering doesn't do the job, remove the trap (see page 17) and flush it. This also lets you get a snake into the main drain.

Toilets

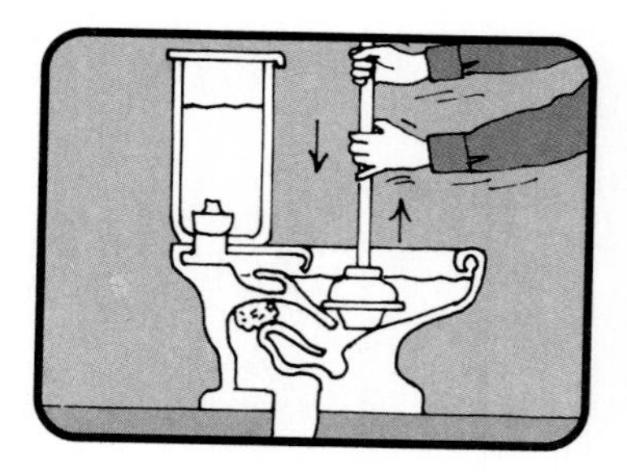

Use a plumber's friend over the hole in the bottom. Work the plunger hard and vigorously, and don't give up too soon.

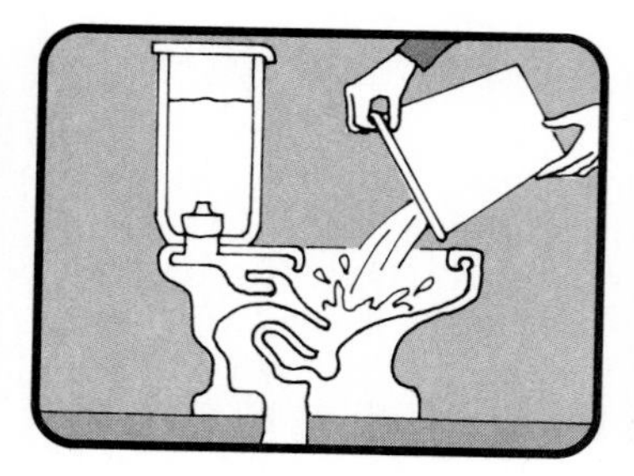

If the toilet doesn't have water in the bowl, fill it to the rim. Spread petroleum jelly on the plunger's rim; this aids suction.

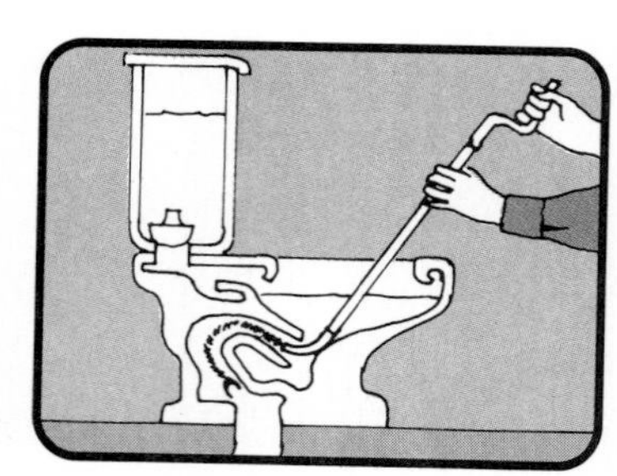

If a plunger doesn't work, use a special snake called a *closet auger.* As you crank, it wends its way through passages.

Tubs

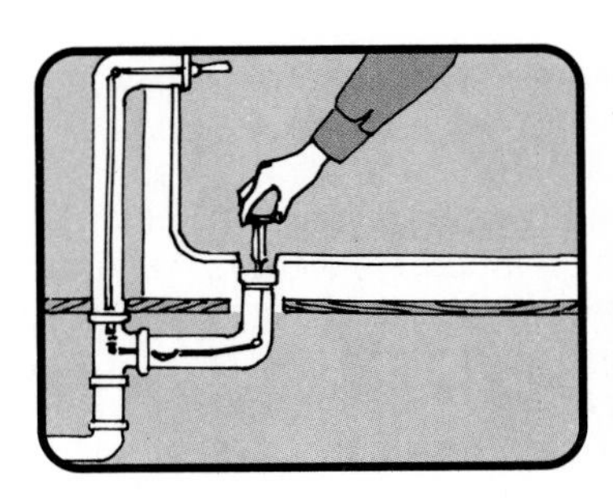

Remove and clean the stopper strainer. Try the plunger treatment, blocking the overflow drain with a wet piece of cloth.

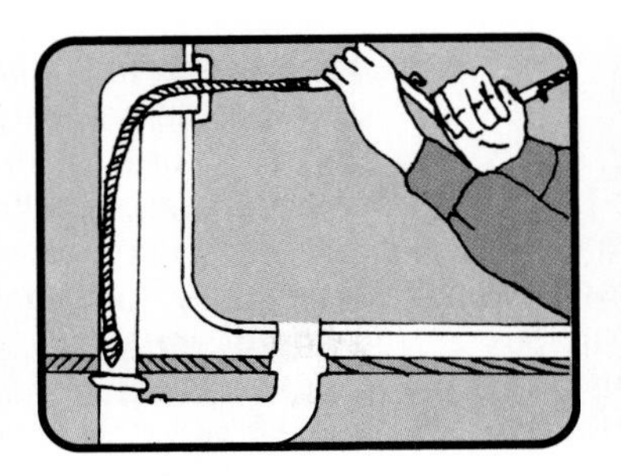

If the plunger doesn't work for you, remove the tub's pop-up or trip-lever assembly and run a snake through the overflow tube.

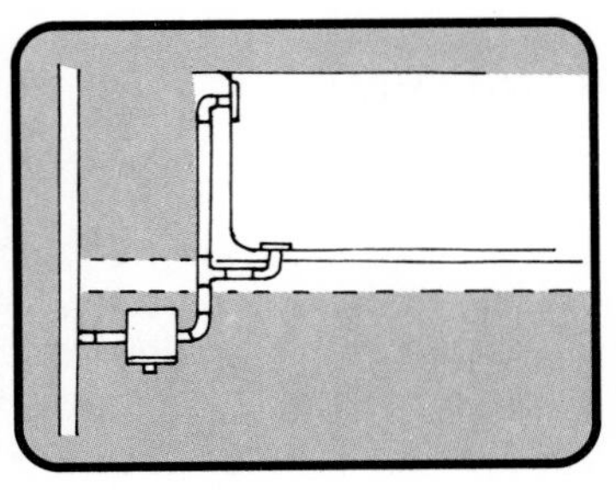

Some tubs have a drum-type trap, accessible by removing a plug that is in or under the floor beside the tub.

Main Drains

There may be more than one cleanout in your home's drain system. If so, find the cleanout plug nearest the sewer line. Clean out this pipe first, then check to see if the stoppage has been removed. If not, move down the line of plugs from the sewer line back.

Sometimes you can save yourself some muscle simply by loosening the cleanout plug. If water drips or forms around the threads of the plug when you loosen it, the trouble is between this plug and the sewer.

With a pipe wrench or adjustable-end wrench, unscrew the clean-out plug. Have a large bucket handy to catch any residue.

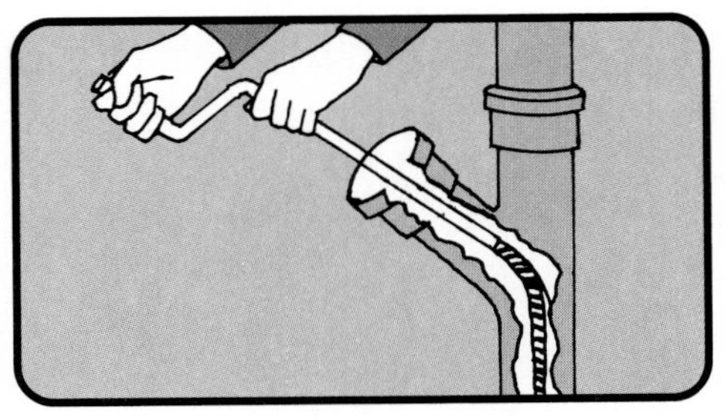

Thread an auger or snake into the cleanout opening and toward the sewer line. Once you break through, flush with a garden hose.

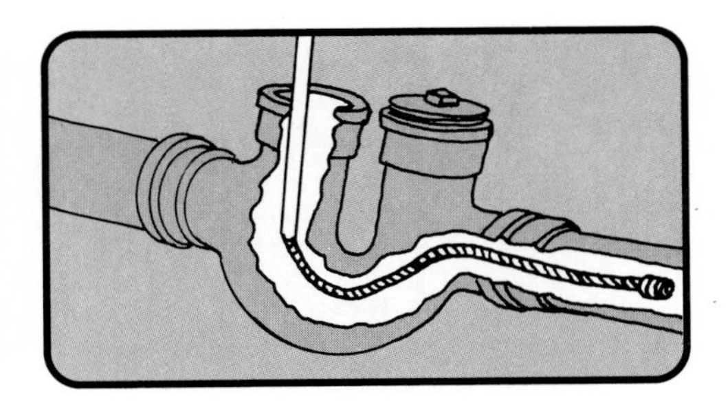

On a U-trap, work from the clean-out plug nearest the sewer line. If the obstruction isn't in the trap, continue up the main drain.

Sewer Drains

The sewer drain, the largest of your home's drains, rarely gets clogged. If it does, though, first remove the cleanout plug. Then thread a garden hose into the line and turn on the water full-blast. Push the hose through the blockage, letting the water pressure clear away the debris. Or insert an auger-type snake into the pipe and twist it through the blockage. If your problem persists, rent an electric auger with flexible blades that snip away the debris, or call in a pro.

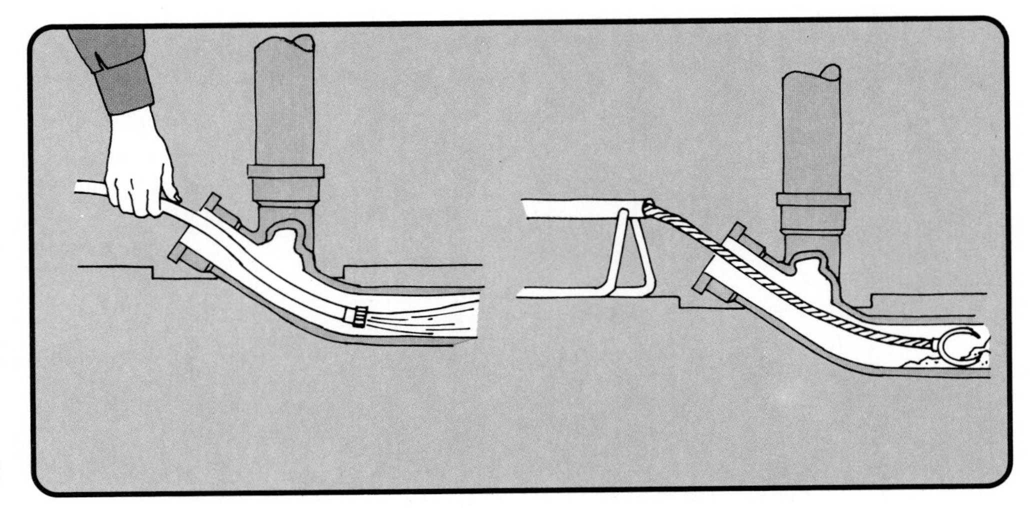

MAKING FAUCET REPAIRS

Unless your ears have long since tuned out such noises, it probably hasn't been too long ago that you heard the plop-plop-plop of a leaky faucet. And you can bet you'll hear it again. No matter how hard you try, wishing away leaky faucets doesn't work. So you might as well learn how to stop that drip.

Though faucets vary considerably in style, all fall into one of two broad categories—*compression* and *noncompression* faucets. Chances are, you've got some of each in your home.

Compression types, also known as *stem* faucets, always have separate hot and cold controls. Turning a handle to its *off* position rotates a threaded stem. A *washer* at the bottom of the stem then compresses into a *seat* to block the flow of water, as shown in the first anatomy drawing below. To learn about repairing stem faucets, see the opposite page.

Most noncompression faucets have a single lever that controls the flow of both hot and cold water. Inside, a single-lever noncompression faucet may have any of four different operating mechanisms, also illustrated here.

Moving the lever on a *tipping-valve* faucet activates a *rocker cam*, which in turn opens spring-loaded valves in the hot and cold water lines. Temperature-blended water then flows through the spout. More about these on page 12.

A *disk* faucet mixes water inside a *cartridge*. At the bottom of the *mixing chamber*, a pair of disks raises and lowers to regulate the volume of water, and rotates to control its temperature. For repair information, see page 12.

A *rotating-ball* faucet consists of a ball with openings that line up with the hot and cold inlets and with the spout. Rocking its lever adjusts both the temperature and the flow. Repairs are explained on page 13.

A *sleeve-cartridge* faucet operates something like a disk type—lifting the handle controls the flow; moving it from side to side regulates the temperature. Inside, though, the workings include a cartridge and sleeve arrangement instead of disks. For repair procedures, see page 13.

Any of these five types of faucets may have aerators, divertors, and strainers that can clog and choke the flow of water. If you have a problem with any of these items, turn to page 15. Note, too, that tipping-valve faucets sometimes have another set of strainers inside their housings, as explained on page 12.

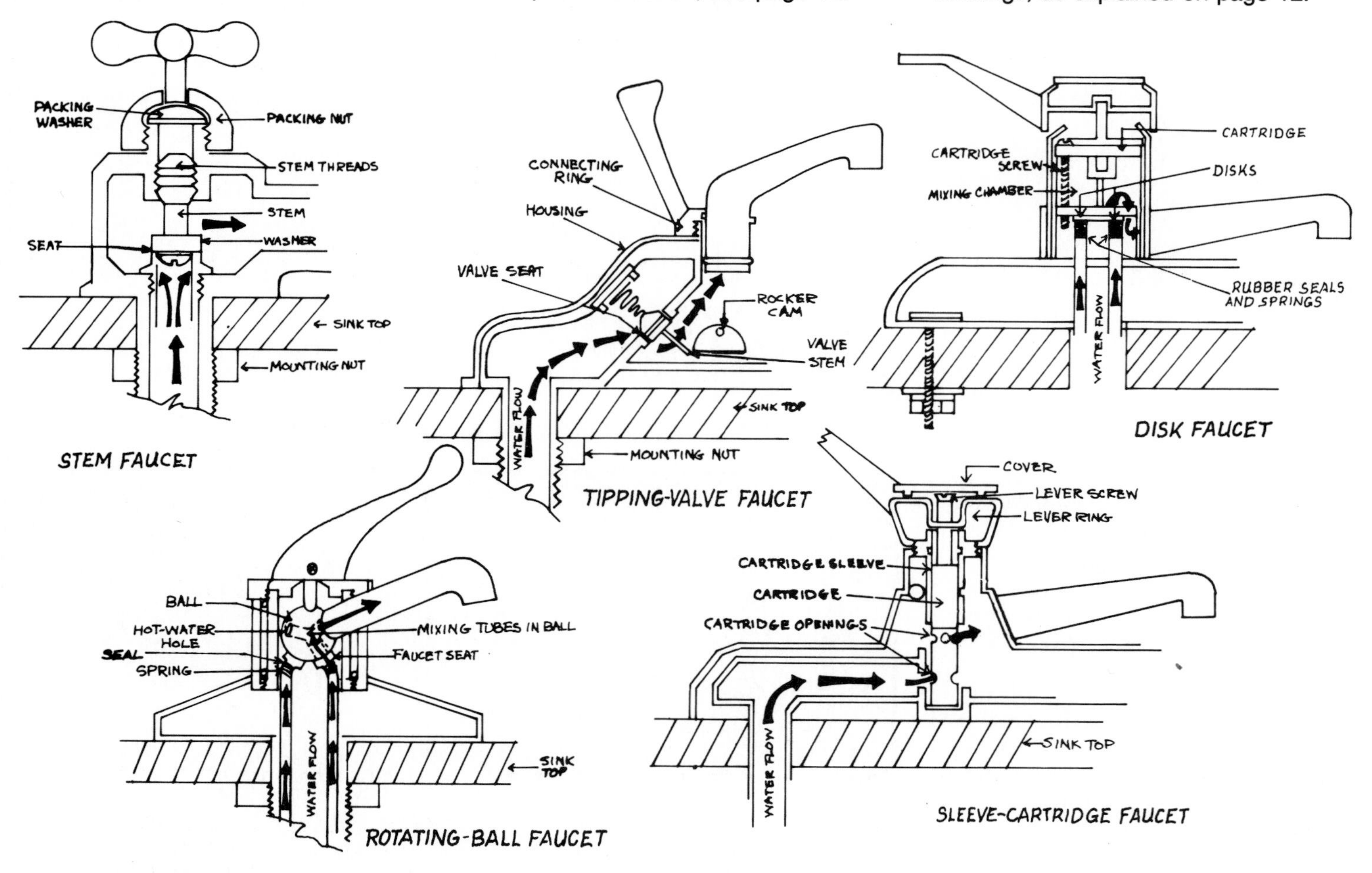

FAUCET ANATOMIES

Repairing a Stem Faucet

Stem faucets—the most leak-prone of all types—develop the sniffles for one or more of the following reasons: a worn washer, a pitted or corroded valve seat, or deteriorated packing.

Examine the anatomy drawing on the opposite page and you'll see that the washer must withstand the pressure of repeated openings and closings. As the washer wears, you have to apply more and more muscle to turn off the water—until finally, no amount of turning can stop the flow.

Fortunately, replacing a washer is a simple procedure, as illustrated below. You'll need only an adjustable wrench, a screwdriver, and a faucet repair kit that includes an assortment of washers to fit most stems. The package generally also includes O-rings and new screws for attaching the washers to the stems.

If you have to replace faucet washers often—every month or two—you're probably dealing with a pitted or corroded valve seat. Abrasion here wears out washers rapidly. Again, the solution is simple, though you'll need one specialized tool—the valve-seating tool or seat grinder, shown below and on page 6.

Washer and seat problems cause drips. If, on the other hand, a faucet is leaking around the handle, its packing has worn out. With newer faucets, which don't use packing, the problem may be a faulty O-ring.

While you have a faucet apart, note whether the threads around its stem show signs of heavy wear. If so, you'll be money ahead to replace the entire unit, preferably with a washerless noncompression type. Installation is fairly easy; see pages 42 and 43.

And whenever you work on a faucet, be sure first to turn off the water at the main entry or at the shutoffs below the sink or lavatory, as shown on page 5. Forget this step and you'll have a real mess on your hands.

Pry out the decorative escutcheon on the faucet handle. Back out the screw and remove the handle. Lift it straight up.

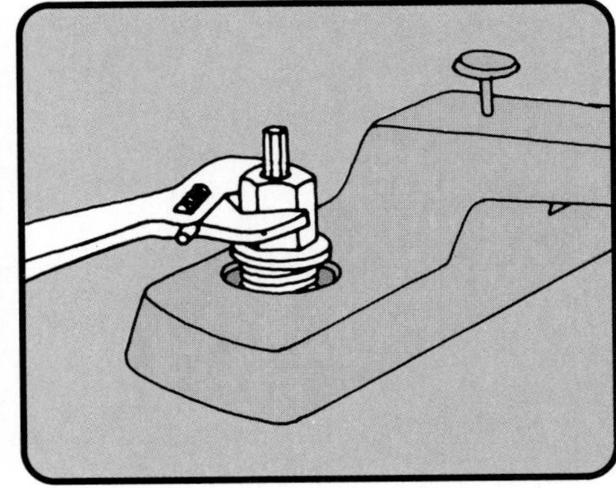

Remove the packing nut. Use an adjustable wrench or a pair of slip-joint pliers for this. Then simply turn out the stem.

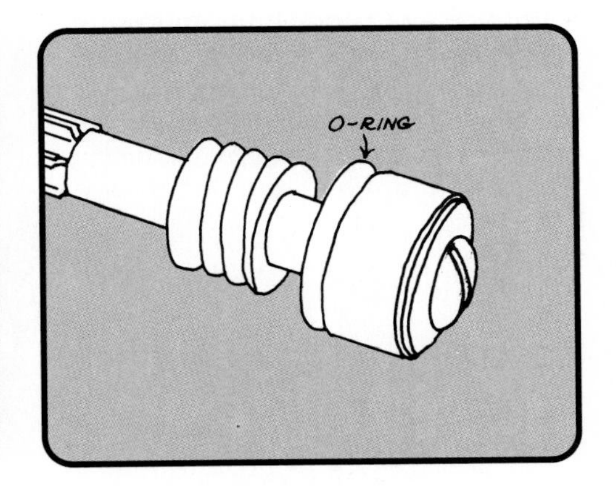

The seat washer, held in place by a screw, is at the bottom of the stem. You may have to replace the stem's O-rings as well.

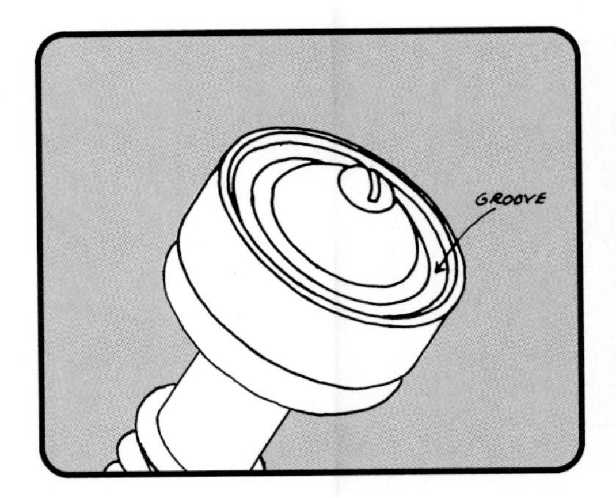

A worn washer will be grooved, pitted, and/or frayed. When you replace it, clean the entire valve stem with fine steel wool.

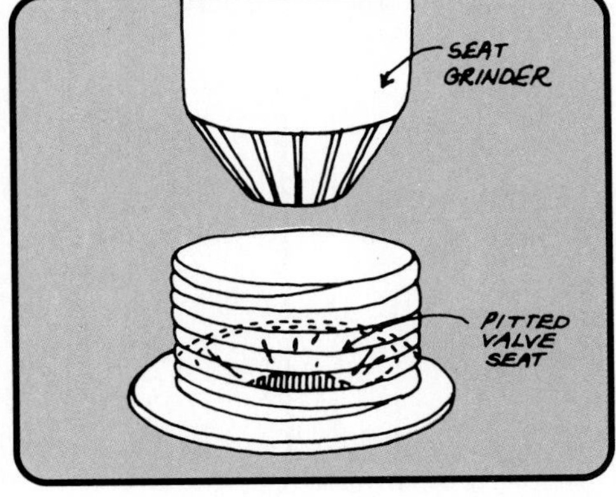

To smooth a worn seat, insert a grinder, apply light pressure, and turn clockwise several revolutions. Blow out chips.

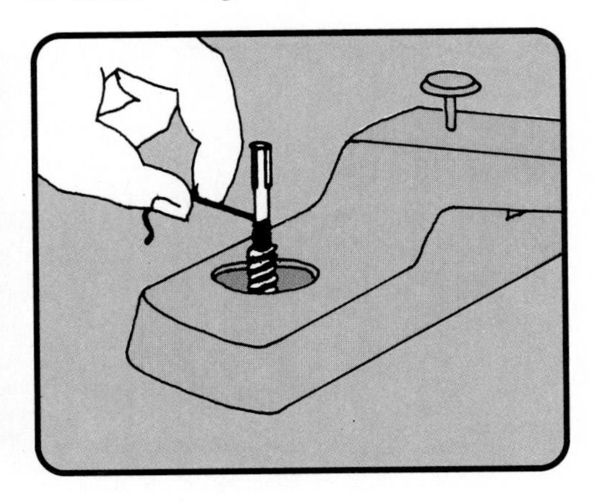

Before you reassemble an older faucet, be sure to wind packing around the stem, then install the packing nut.

Repairing a Tipping-Valve Faucet

Unlike compression faucets, tipping-valve units have no washers—but worn valve-seat assemblies can cause them to drip anyway. To get at these, you remove the spout and housing, as shown. Then replace the assemblies—valves, seats, seals, and all—with parts sold in kit form. With some, you may need a seat wrench to get the seat out.

Sometimes sediment can gum up strainers in the valve assemblies, making it seem as if there's something lacking in your water pressure. Rinse out the strainers and water will flow again.

When water oozes up from the spout's base, the O-ring has probably gone bad. Take care in removing it that you don't scar the metal underneath. Lubricating with petroleum jelly helps you slip on a new O-ring and makes a better seal.

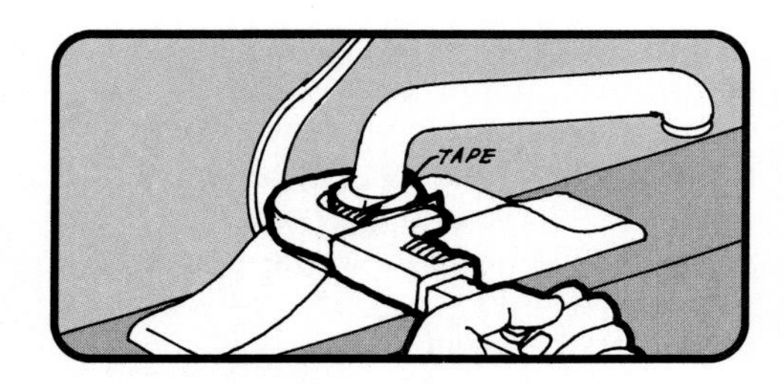

First turn off the water supply, then pad the jaws of a wrench with tape and loosen the nut at the spout's base.

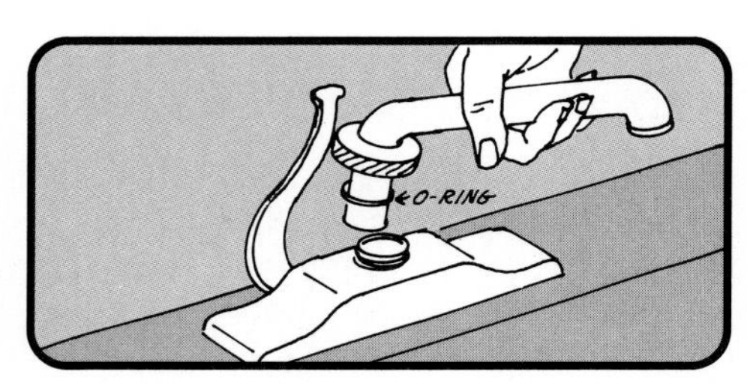

Now lift the spout straight up and out, taking care not to damage the O-ring. Lift or pry off the housing, too.

Plugs on either side—one for hot, one for cold—secure the valve-strainer assemblies. Remove these and pull out the parts.

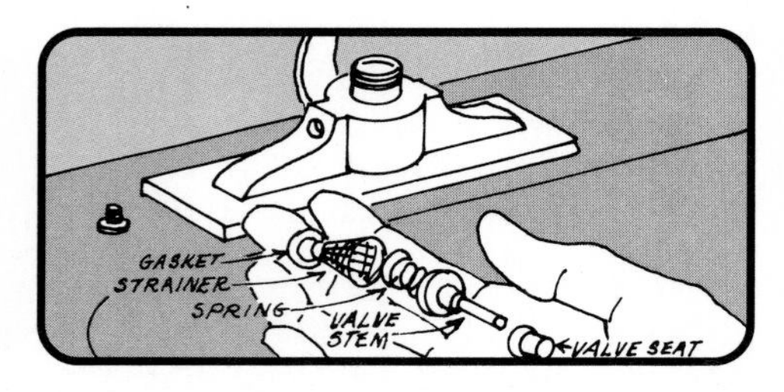

Each assembly consists of a gasket, strainer, spring, stem, and seat. Take these along when you shop for replacements.

Repairing a Disk Faucet

The disks in most disk faucets are made of ceramic material and won't wear out. But their inlet holes (see below) can become constricted by lime deposits in the water. When this happens, you have to dismantle the faucet and clean out the debris. If the faucet leaks at its base, you must replace the inlet seals in the cartridge's underside.

Another type of disk faucet (not shown) resembles the compression versions shown on pages 10 and 11 in that it has separate hot and cold controls. Take one apart, though, and you'll find a cone-shaped rubber diaphragm at the end of the stem where you'd expect to see a washer. If this is worn, pry it out and replace it. These are sometimes called "washerless" faucets.

You can buy repair kits for both ceramic and diaphragm-disk faucets, but take along the old assembly when you shop; sizes vary. As with any faucet repair, shut off the water and drain the tap before you begin.

Pry off the decorative cap, then remove the screw and handle. With some, you pry as indicated to get at the handle screw.

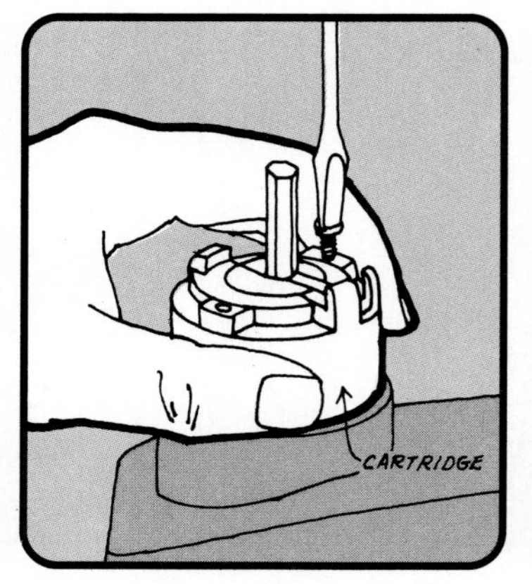

Now remove the two screws that hold the cartridge in place and lift it out. This entire unit can be replaced if necessary.

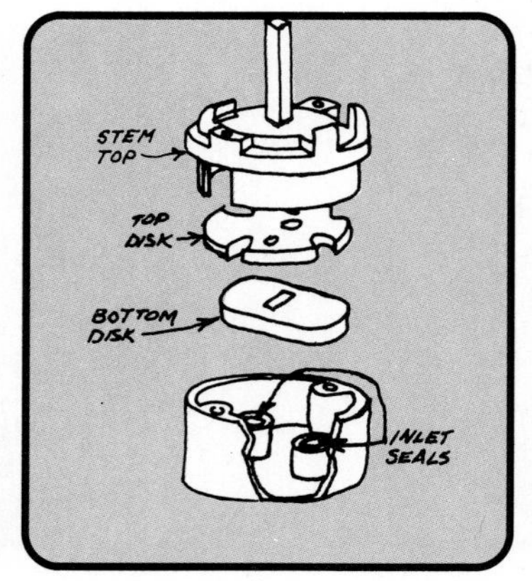

But first check to be sure that dirt hasn't lodged between the disks, and that the inlet seals are in good condition.

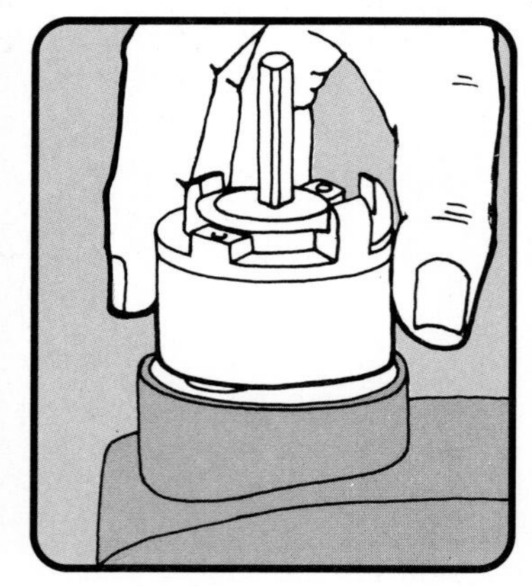

Insert the new or repaired assembly back in place, carefully aligning it so the screws mesh with holes below.

Repairing a Rotating-Ball Faucet

Ball faucets serve for years without trouble. When one begins to drip, you can be almost certain its springs and seats need replacing. Leaking around the handle, on the other hand, means it needs new O-rings. Neither of these repairs is a difficult job.

The procedure for getting at faucet parts varies somewhat, depending mainly on whether the unit has a fixed or a swiveling spout. With fixed-spout models, you simply remove the handle and a cap underneath, as shown here; with swivel-spout types, you have to lift off the spout as well.

While you have the unit apart, check the ball itself for wear or corrosion and replace it, too, if necessary. Repair kits include springs, seats, O-rings, and other seals—but you'll need the make and model number or the old parts to get the right components.

Reassemble the parts in order and replace them in the housing. With a swivel-spout model, push the spout straight down until you hear it click against a slip ring at the base of the housing. Since the O-rings create lots of tension, you'll have to push hard.

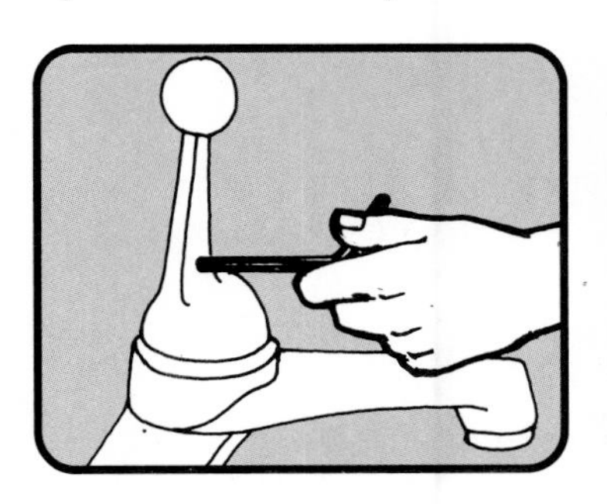

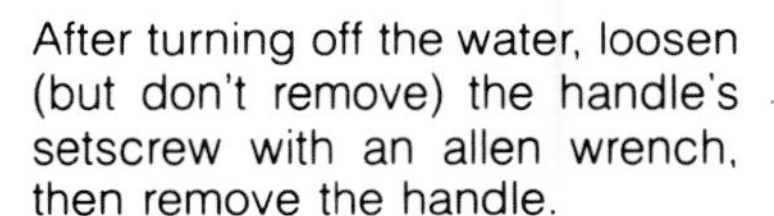
After turning off the water, loosen (but don't remove) the handle's setscrew with an allen wrench, then remove the handle.

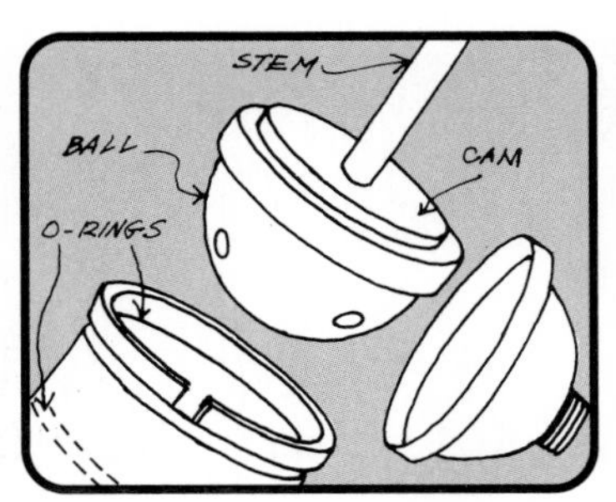

The cap comes off next, then pull out the cam, ball, and stem assembly. Remove the O-rings if they're worn or cracked.

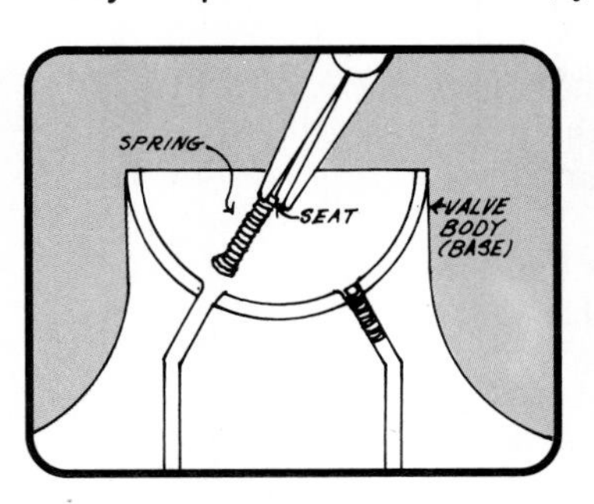

Pull out the seats and springs with long-nose pliers. Replace these according to directions that come with the repair kit.

When you replace the ball, be sure to align a slot in its side with a pin inside the housing. The cam has a lug key, too.

Replacing a Sleeve Cartridge

When a sleeve-cartridge faucet goes bad, you'll have to replace either its O-rings (if there are any), or the entire cartridge. These assemblies don't lend themselves to repairs—but they're not prohibitively expensive.

The key to dismantling one lies with a small "keeper" or retainer clip at the base of the handle assembly. With some faucets, you can see this clip at the point where the handle meets the base. With others, you must first remove the handle and—in the case of a swivel-spout faucet—the spout.

Under the handle, you'll probably find a ring or tube that simply slides off to expose the keeper. Pry out the clip with a screwdriver or long-nose pliers and the cartridge will pull out with little difficulty.

When you assemble the faucet, look for a flat spot, arrow, or other mark on the cartridge stem. Usually this must be pointing up for the faucet to work properly.

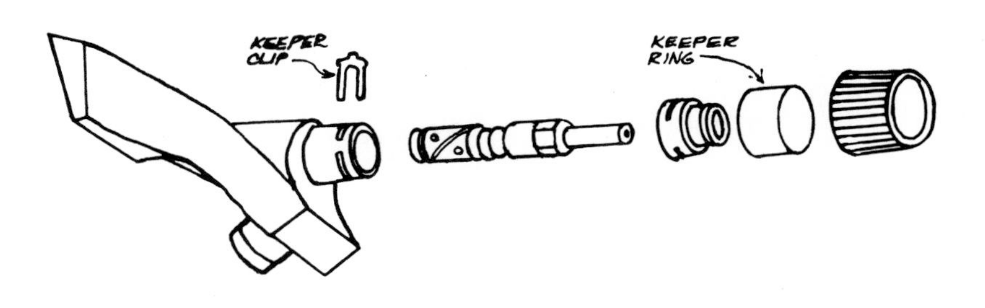

A typical cartridge faucet looks like this. A keeper ring must be pulled back to expose the keeper clip for removal.

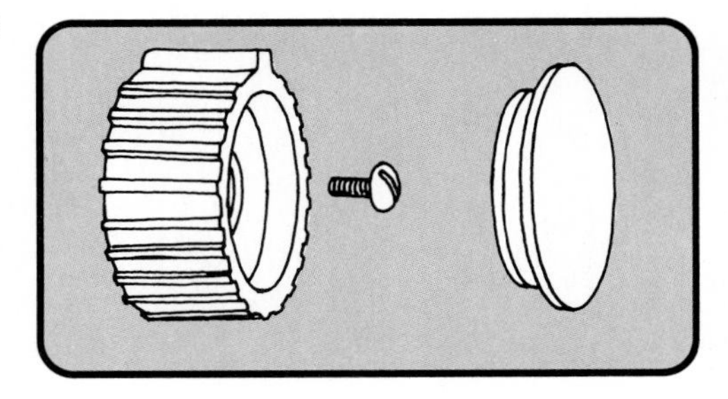
With a screwdriver, carefully pry off any decorative cap, and back out the screw in the handle. Then withdraw the cartridge.

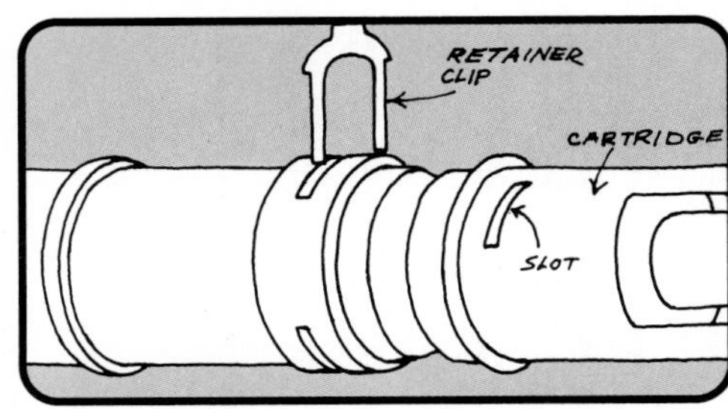

The clip may not be hidden at all. Some faucets also have a second retainer clip located near the handle.

REPAIRING TUB/SHOWER FAUCETS

Like sink and lavatory faucets, wall-mounted faucets also fall into two categories: compression and non-compression types. The non-compression version usually has a single handle pull-on, push-off configuration, with a cartridge assembly beneath. When it leaks, this assembly usually requires replacement. Compression faucets, the two-handled types, feature O-rings and washers, which you can replace. An example of each type is shown here.

Problems with shower heads usually stem from lime deposits and/or corrosion. Often, you can disassemble the head and clean its screens and strainers. If you can't take your unit apart, replace it with a new one.

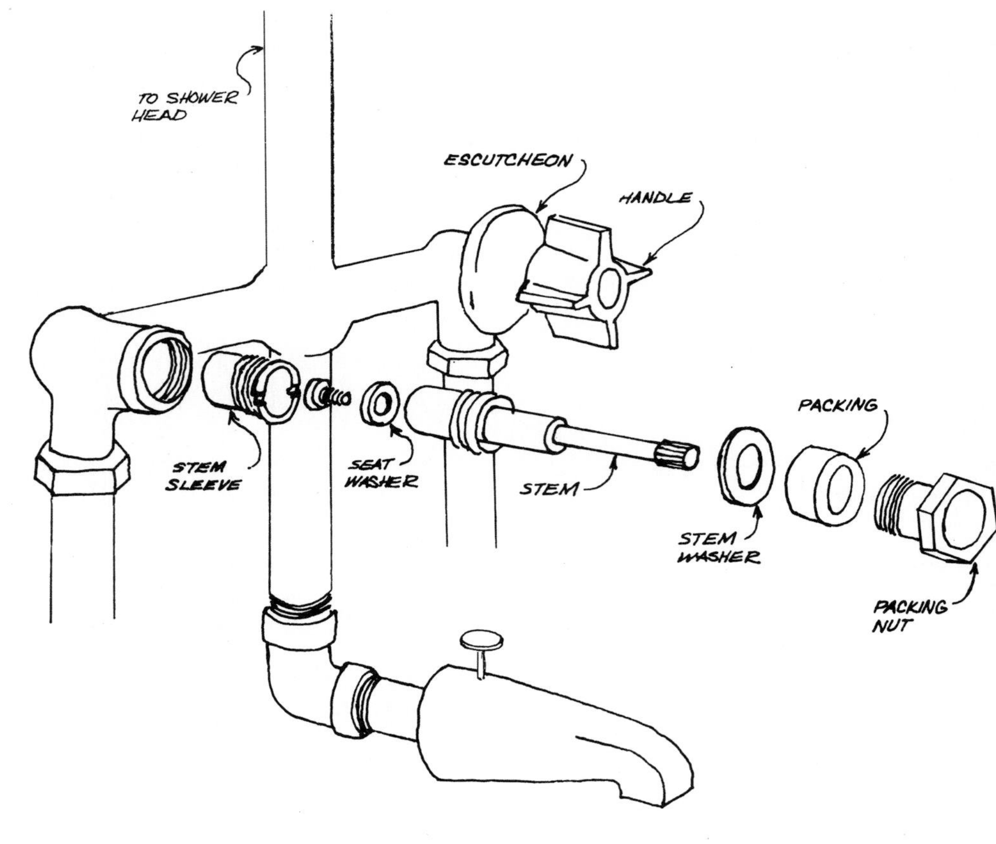

COMPRESSION-TYPE FAUCET

To repair compression faucets, pry out the handle insert and remove the knob. Under this you'll find a packing nut. Loosen this nut and replace the handle on the stem. Then, turn the handle to remove the stem.

If the stem has worn O-rings or a worn seat washer, which you'll find at the bottom of the stem, replace them. Thinly coat the O-rings with a heat-resistant lubricant jelly. Replace worn packing, too, if your stems have it.

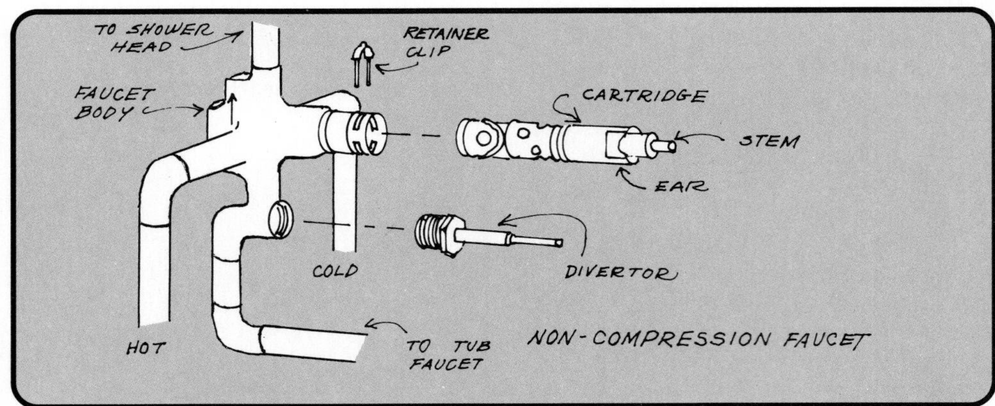

Unscrew shower heads as shown in this sketch. This will expose the screens and strainers for cleaning or replacement.

To repair leaky single-handle faucets, you need to replace the cartridge. Remove the handle and retainer clip, then the cartridge.

To insert the new cartridge, push it into the housing until the ears are flush with the housing. Align parts, then insert clip.

REPAIRING DIVERTORS, SPRAYS, AND AERATORS

Whenever divertors, sprays, or aerators act up, you'll usually find the culprit to be a worn washer or a clogged strainer.

Divertors channel water from a faucet to a shower head or spray attachment. You can make minor repairs such as replacing worn O-rings, packing, or washers by backing out the divertor assembly. However, if the divertor assembly is leaking, you'll have to replace it. Take the old unit to a plumbing shop so you can match it with a new one.

Sprays have a hose and nozzle head. Troubles can develop in the connections, washers, or the nozzle. But before you rip into the assembly, try tightening connecting nuts to stop leaks, and make sure the hose is not kinked.

Aerators, those tiny spray devices connected to the spouts of faucets in sinks and lavatories, have threads that let you screw them to the spout. Corrosion in the form of rust or lime deposits blocking the screen or strainers causes most of the problems you'll encounter with these devices. If a malfunctioning aerator is an old one, replace it with a new assembly.

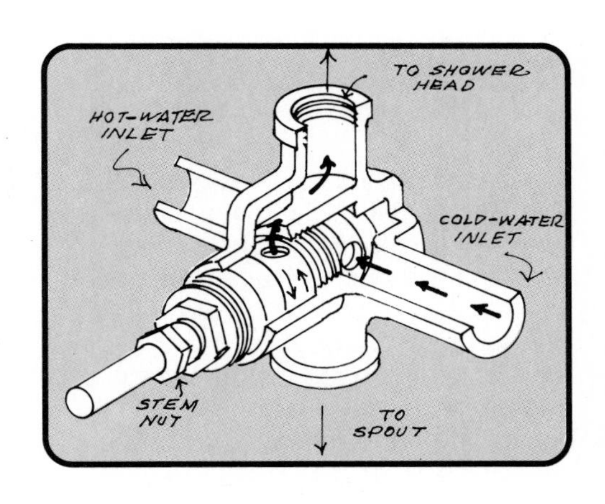

Remove the faucet handle and unscrew the stem nut to release the innards of a shower divertor assembly.

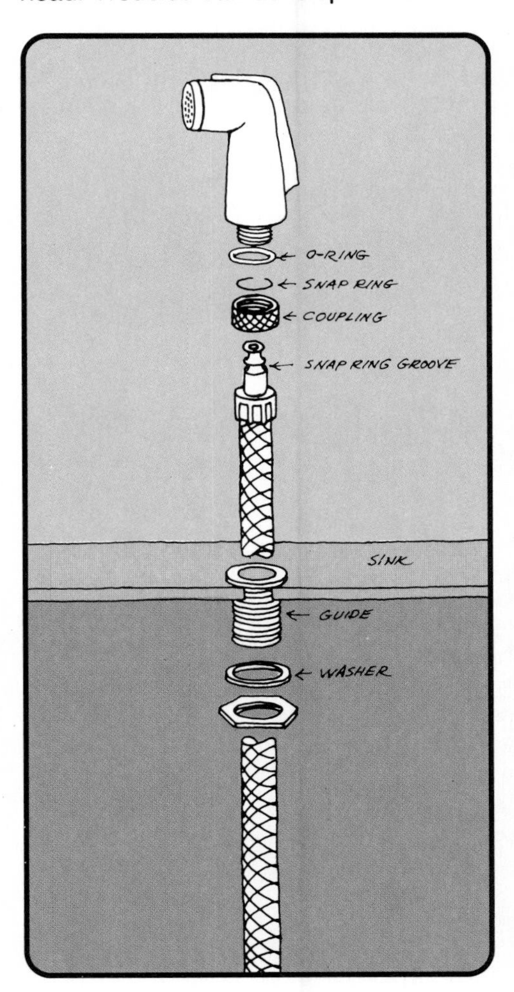

A spray attachment has washers and couplings that can leak. You can stop most leaks by tightening these.

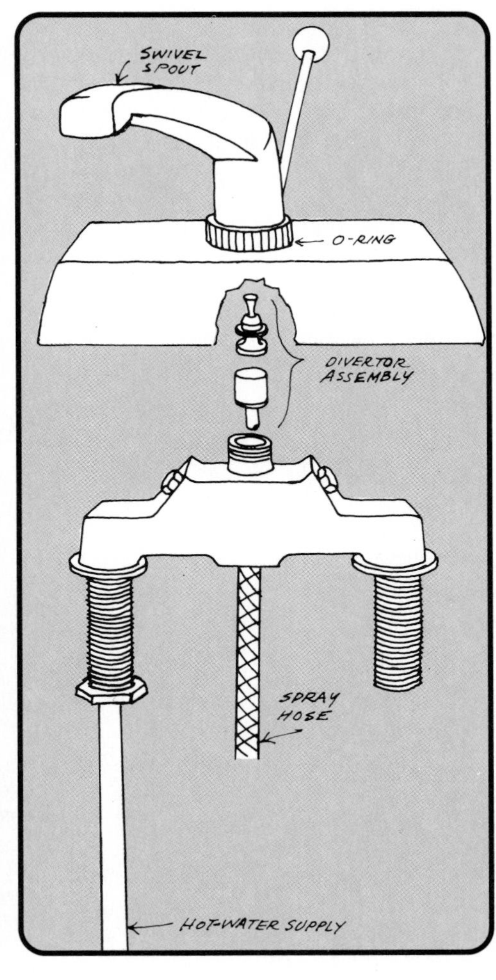

The divertor on sink faucets is positioned on top of the faucet housing. The spray hose connection is under it.

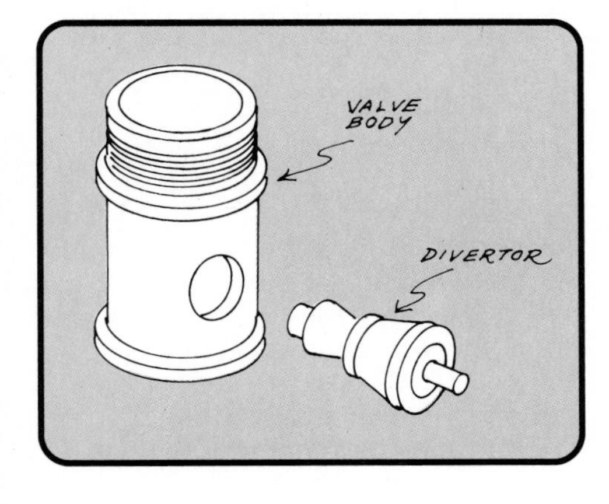

Lime deposits can clog a faucet's divertor. With the type shown, remove and replace the assembly. You can clean some kinds.

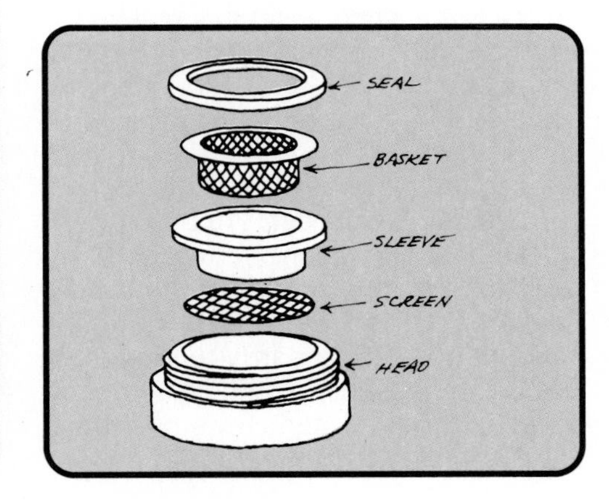

Assemble aerators in this order. You can clean the basket and screen by flushing them with water or brushing the mesh.

REPAIRING DRAINS AND TRAPS

Compared to faucets, the components that carry water away from a fixture are relatively simple. Since gravity does most of the work, drainage systems need to handle only moderate pressures. And except for the rudimentary linkages that operate pop-up and trip-lever assemblies (see below), you'll find no moving parts underneath a lavatory, sink, or tub.

When trouble does occur, it almost always falls into one of two categories—a clog or a leak. If the flow seems just a little slower than it should be, try running hot water down the drain for about 10 minutes; often this will be enough to dissolve any accumulated grease or soap.

If hot water doesn't do the job, try a commercial drain cleaner. Follow the manufacturer's instructions to the letter, though. And don't use drain cleaner in a completely stopped-up fixture—if it doesn't work and you have to dismantle the trap, as shown opposite, you'll be working with dangerously caustic water. For more about clearing clogged drains, see pages 8 and 9.

Leaks call for some sleuthing to pinpoint exactly where the water is coming from. First set up a bright light under the fixture and wipe all drainage components dry. Now run tepid water (cold water can cause misleading condensation) and methodically check each connection, starting up top where the drain exits the fixture. Don't rely on your eyes alone. Instead, wipe each fitting with your fingertips, then look for moisture on them.

If you find a leak at a connection, often simply tightening its slip nut will solve the problem. Take care not to apply too much pressure, though. Because drain fittings are made of soft, lightweight materials, they can be easily cracked or crushed.

If tightening doesn't do the job, prepare to dismantle the assembly, as shown opposite.

Adjusting Pop-Ups and Trip-Levers

Have you ever filled a lavatory or tub, stepped away for a few minutes, and returned to discover that the water level had dropped? If so, the fixture's pop-up or trip-lever mechanism is letting you down.

Start by lifting or turning out the *stopper* and flushing away any hair, soap, or other debris that might be preventing it from seating properly. Wipe off the *flange,* too, and inspect it for any signs of wear or damage.

After you replace the stopper, check to see if the lifting assembly pulls it down snug. If not—and you're dealing with a lavatory pop-up—get under the basin and take a look at the *pivot rod.* It should slope slightly upward from the *pivot* to the *clevis.*

To adjust this, loosen the clevis' setscrew, push the stopper down hard, and retighten the setscrew. Now the lavatory will probably hold water, but its *lift rod* may not operate as easily as before. If this is the case, adjust the linkage between the pivot and clevis so they meet at nearly a right angle.

Occasionally a sink pop-up will leak at its pivot. Sometimes tightening the retaining nut here will stop the drip. If not, remove the nut and replace any washer or gasket you find underneath.

Tub pop-ups—and a variation called a trip-lever—work in much the same way, except that they are housed within the overflow tube. To adjust them, you remove the stopper, unscrew the overflow plate, and pull out the entire assembly.

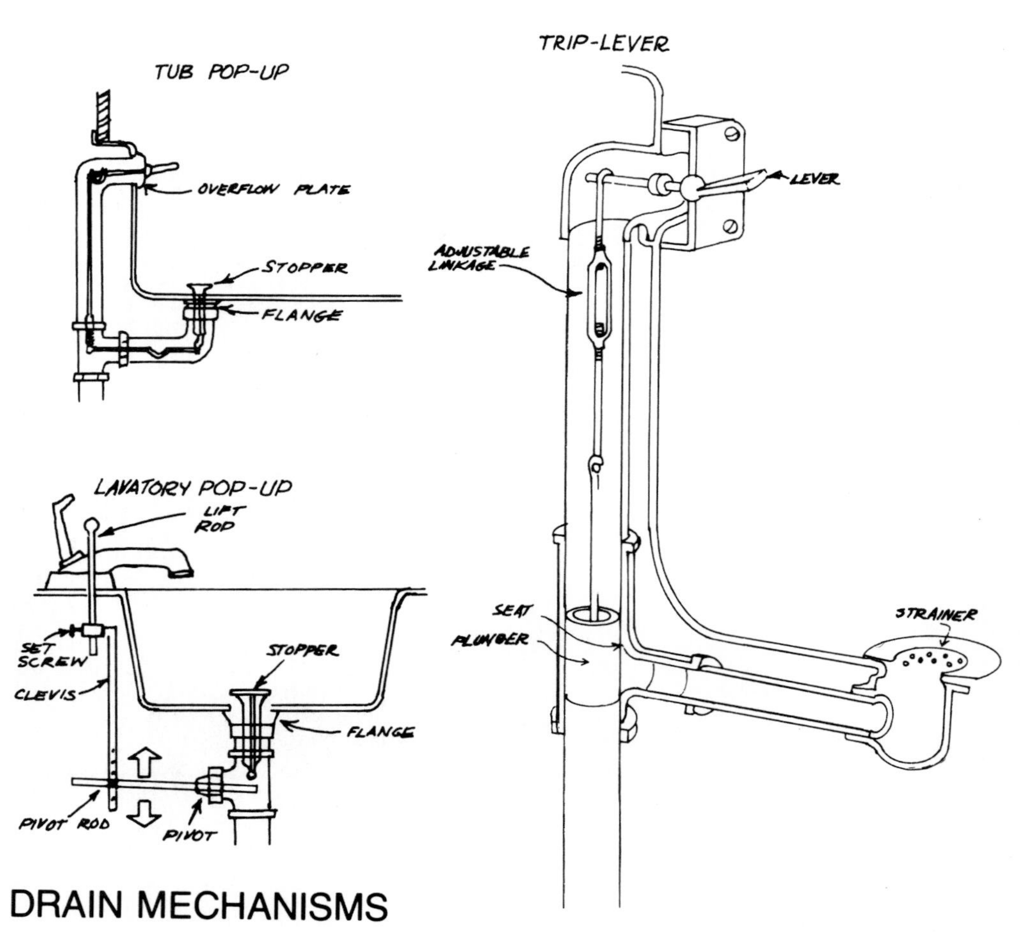

DRAIN MECHANISMS

Dismantling a Trap

Clogged drains, missing rings, and the ravages of time make it almost inevitable that you'll have to take apart a trap at some time or other. Master the steps depicted here, though, and you'll be out from under the problem in just a few minutes, and with a minimum of aggravation.

The secret to the way trap components fit together lies with the special slip-joint connections depicted below. Only the trap itself is threaded, not the tailpiece or drainpipe that slips into either end. This arrangement lets you twist everything around to align the assembly. Then you tighten slip nuts to secure it.

Although most slip fittings utilize rubber washers, some older slip fittings may be packed with lamp wick, which looks like ordinary cotton string, but makes a more watertight seal.

Whenever you have a trap apart, inspect it carefully for signs of corrosion. These usually show up first at the bend along the sides and bottom. Before going shopping for a new trap, measure the diameter of the tailpiece. This is usually 1¼ inches.

First shut off water at the fixture stops or take knobs off the faucets. Don't chance inadvertently turning them on.

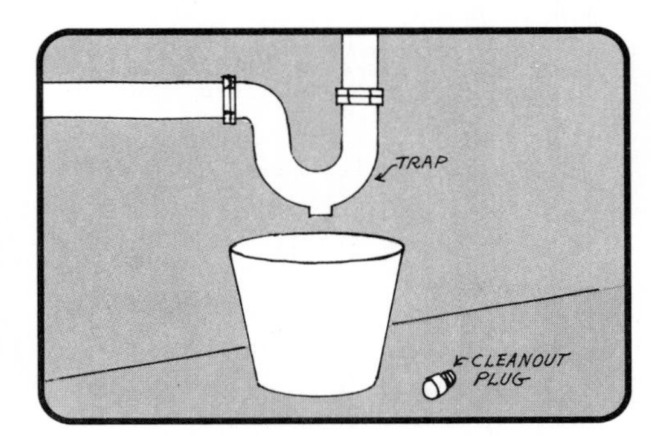

Next, slip a bucket or tray underneath to catch water in the trap. Open the cleanout if the trap you're working with has one.

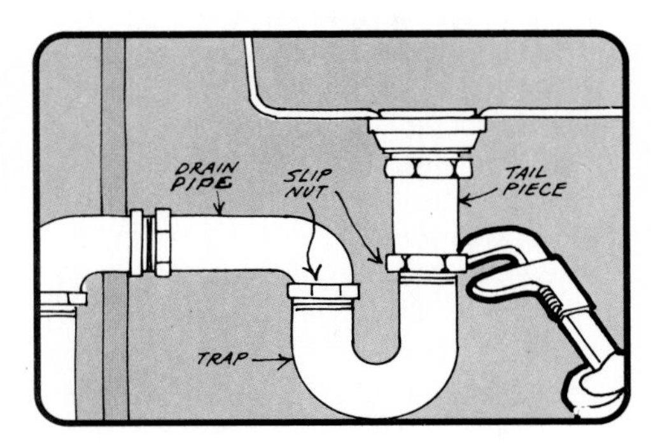

When you loosen slip nuts at the tailpiece and drainpipe, this type of trap will simply drop loose or come off with a tug.

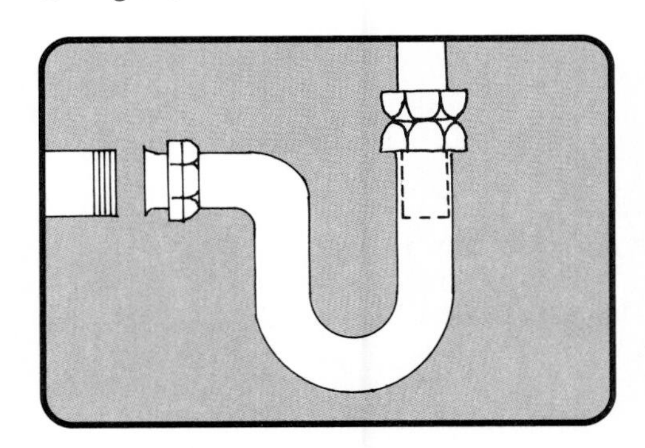

With a fixed trap, such as this one, you slide down the tailpiece, then turn the trap loose from the drainpipe.

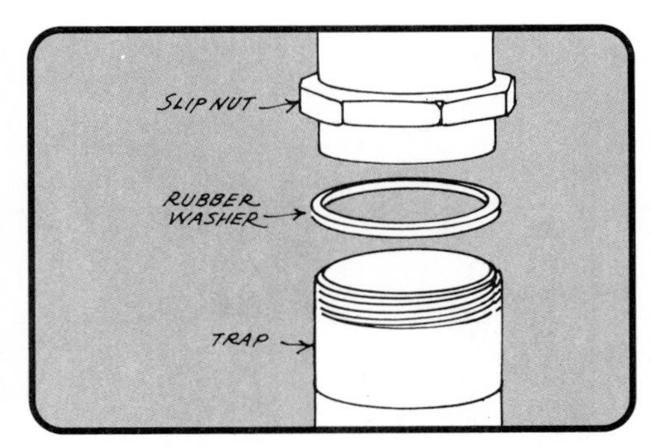

Leaks often result from worn-out washers rather than from the trap itself. Tighten the slip nut or replace the washer.

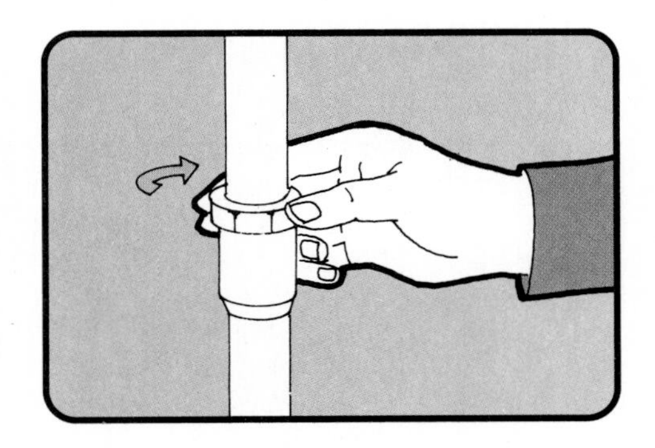

When you reassemble drain fittings, be careful not to overtighten them. Start by turning the slip nuts hand-tight.

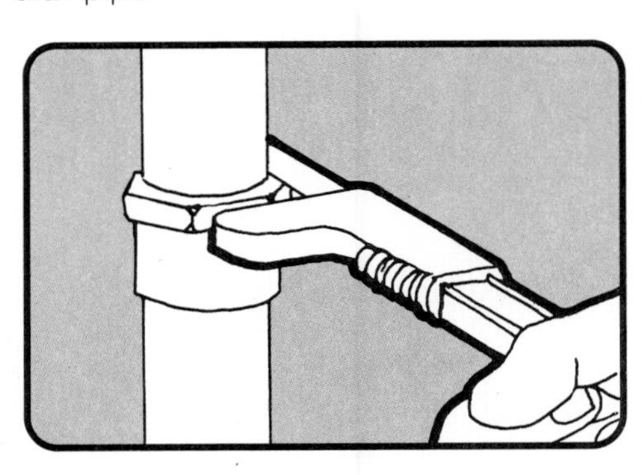

Then go another quarter-turn with a wrench. Select one with smooth jaws—or pad serrated jaws so they don't mar the plating.

To test a trap for leaks, completely fill the basin, then open the drain and check all connections closely for signs of moisture.

REPAIRING TOILETS

Repairing a problem toilet isn't anyone's idea of a good time, but you'll have to do it every so often nonetheless. Most toilet maladies happen inside the tank where all the mechanical parts are located. Only rarely will other problems develop.

Lift the top off a home toilet tank and you'll find—mostly submerged in water—an assortment of balls, tubes, and levers similar to those illustrated at right. To understand what they do, first realize that flipping the handle sets in motion a chain of events that releases water to the bowl, then automatically refills both the tank and the bowl.

In the flushing cycle, moving the *handle* activates a *trip lever* that lifts a *flush ball* at the bottom of the tank. Water then rushes through a *seat* into the toilet bowl. After the tank empties, the flush ball drops back into its seat.

Flushing also triggers the refill cycle, thanks to a *float ball* that goes down along with the water level and opens an *inlet valve*. This brings fresh water into the tank via a refill tube; it also sends water to the bowl through a second refill tube that empties into an *overflow tube.*

As the water rises, so does the float ball. When it reaches a point ¾ inch or so below the top of the overflow, the float shuts off the inlet valve.

The following pages take you step by step through just about everything that can possibly go wrong inside a toilet tank, and tell you what to do about each situation. To learn about toilet bowls and how to install a new unit, see pages 44 and 45.

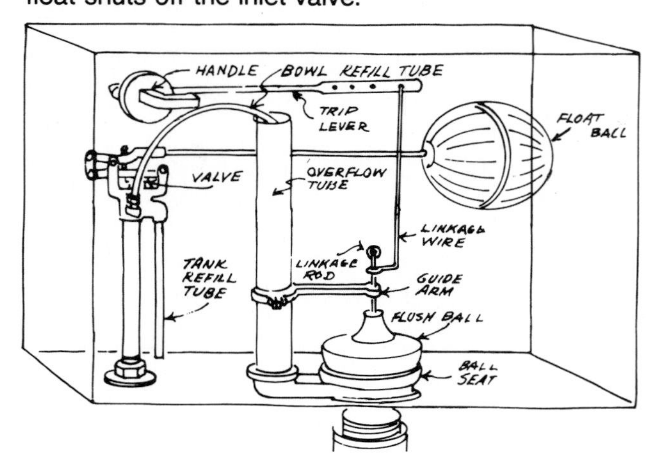

ANATOMY OF A TOILET TANK

TROUBLESHOOTING TOILETS

PROBLEM	SOLUTION
Toilet won't flush	Check the handle, trip lever, guide arm, flush ball, and the connections between each one of the parts to make sure all are functioning. The handle may be too loose or tight; the trip lever or guide arm may be bent or broken; the connection between the trip lever and guide arm may be broken or out of adjustment so it doesn't raise the flush ball far enough.
Water runs, but tank won't fill properly	The handle and trip assembly may be malfunctioning. See above. Check the flush ball for proper seating; check the seat for corrosion; and check the float ball for water inside.
Water runs constantly after the tank is filled	You may have to adjust the float ball downward. Check the float ball to make sure it's not damaged. It could be full of water, causing it to float improperly. The inlet valve washers may be leaking and need replacement. Check to see that the flush ball is seating properly. Check the ball seat for corrosion.
The water level is set too high or too low	Gently bend the flush tank float downward to lower the water level. Bend it upward to raise the water level. Or, use the adjustment screw on top of the inlet valve to set the float arm. The water should be ¾ inch below the top of the overflow tube.
Toilet won't flush properly	Water may be too low in the tank. If so, bend the float ball up to permit sufficient water to flow into the toilet bowl.
Water splashes in the tank while it refills	Adjust the refill tube that runs into the overflow tube. You may need to replace the washers in the inlet valve.
Tank leaks at the bottom	Tighten all the nuts at the bottom of the tank. If this doesn't work, replace the washers.

Repairing Flush Mechanisms

An occasional gurgle, a constant flow of water—both are sure signs that your toilet's got troubles. But don't panic. Almost 100 percent of the time, you can trace the problem to the flushing mechanism, which controls the water in the flush tank. And, usually, correcting the problem involves only a very simple adjustment, or, in some cases, a few new parts.

If the working parts are metal (usually brass) and aren't too old, you're best off replacing the individual malfunctioning parts. But if the parts are plastic or have been in service for several years, replace the entire assembly. These come in kit form with easy-to-follow installation instructions.

Not covered here are trip arms and linkage wires. These parts often corrode and break before the inlet valves, refill tube, and flush-ball seat go on the fritz. To learn about trip arms and linkages, see page 21.

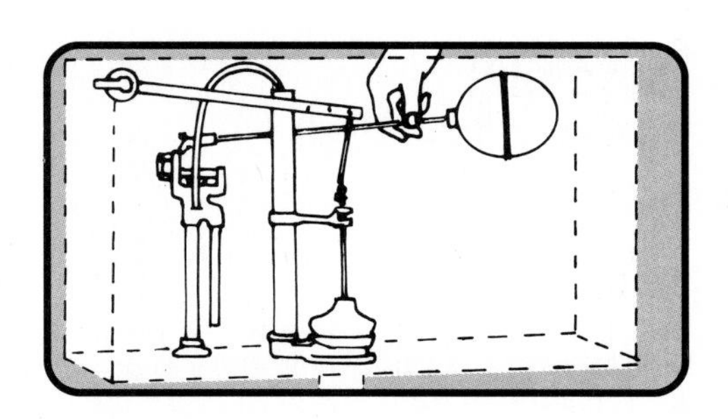

Lift the float rod gently. If the water shuts off, the float ball position has to be changed slightly to close the valve.

To adjust the position, carefully bend the float rod so the float ball is about ½ inch lower. Flush the tank to check the float.

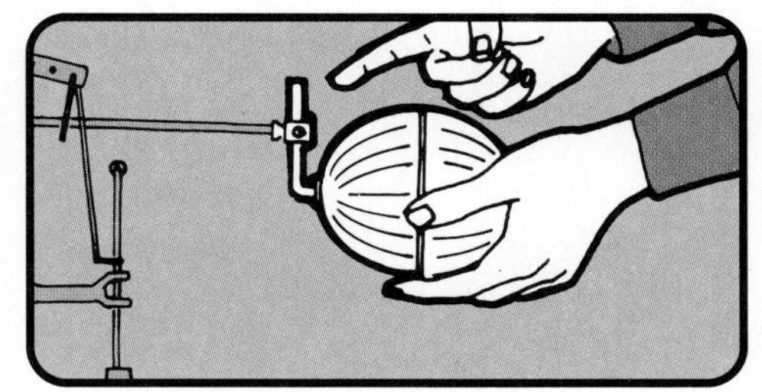

A collar you can buy allows more accuracy when adjusting float position. Also look for an adjustment screw on the inlet valve.

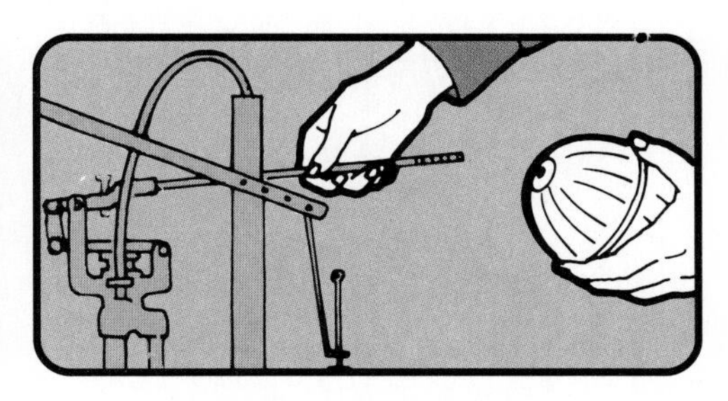

Only half of the float ball should be submerged. If it sinks lower, check it for leaks. Replace the float, if necessary.

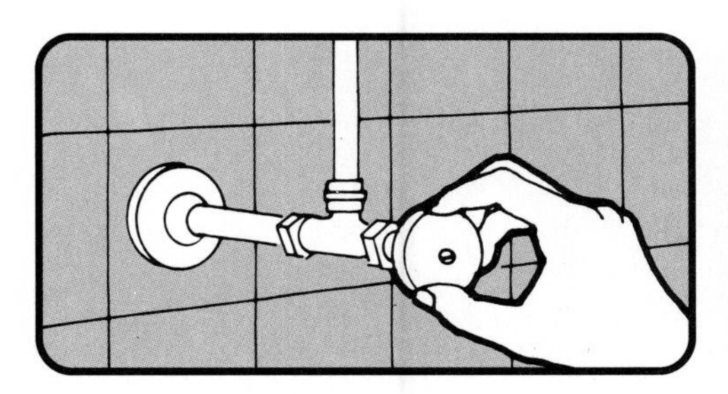

Before making valve-assembly repairs, turn off the water at the shutoff valve below the tank (or at the water meter).

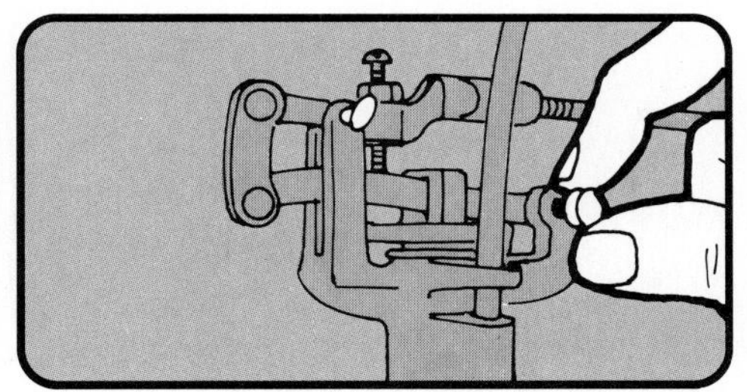

If the water doesn't shut off when you lift the float, the washers may be worn. To open the valve, remove two pivot screws.

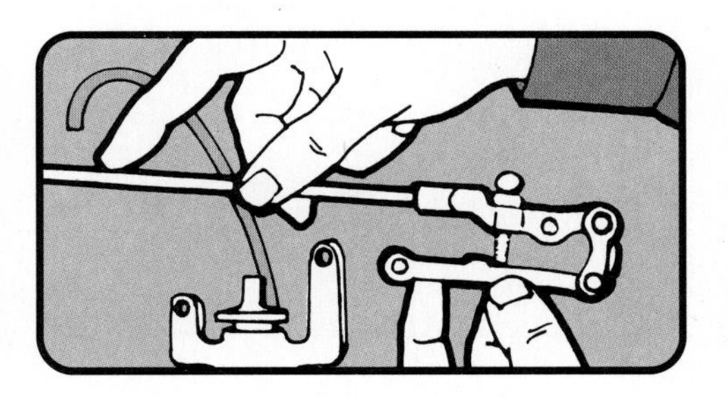

Slide the float, rod, and linkage out of the valve. On some assemblies, you remove a cap that covers the inlet valve.

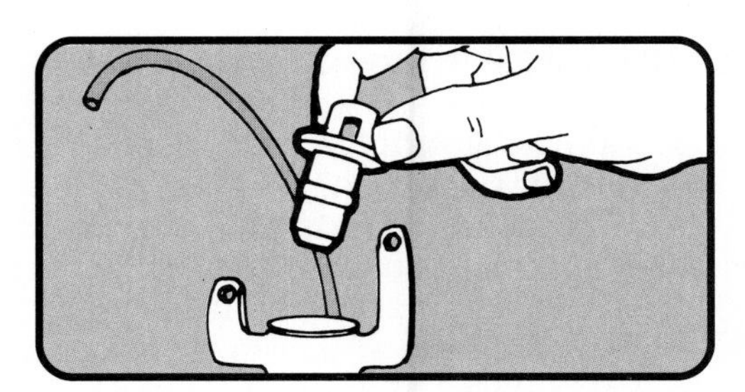

To remove the plunger from the valve, pull upward. If it's stuck, use a screwdriver to gently pry it out. Don't damage the metal.

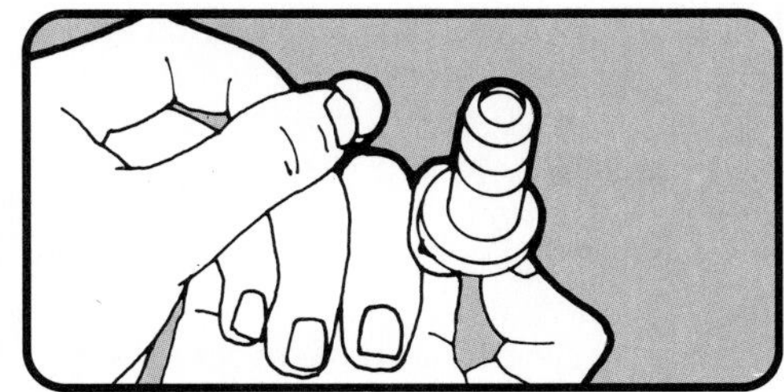

A washer at the base of the plunger shuts off the flow of water. In most cases, you simply push the new washer in position.

Some plungers have two washers. The second washer fits into a groove in the valve. Remove any corrosion from the plunger.

Repairing a Leaky Flush Ball

Generally, you'll find one or more of the following conditions responsible for water leaking via the flush ball from the tank into the bowl: a misaligned guide arm and wire, a bent linkage wire, a worn flush ball, or a pitted or corroded flush-ball seat.

You generally can pinpoint the problem by flushing the toilet and watching these parts operate. (See page 18 for a sketch that identifies them.)

To check the ball seat, lift the ball with the flush handle and run your fingers over the seat. If it feels rough, chances are good that it's corroded or pitted—a job for an abrasive or steel wool.

If your problem is a worn flush ball, buy a replacement. Just unscrew the old ball and screw on the new one. You might want to upgrade the assembly at this time with a new flapper-type ball unit or a "water saver" valve-and-ball device. Both are easy-to-install replacements you'll find at most home center stores. More about these at the bottom of the page.

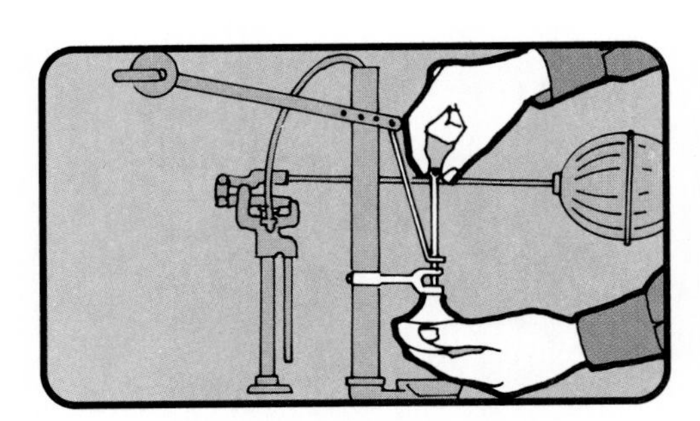

Raise the linkage rod and test the flush ball for wear. To replace the ball, unscrew the linkage rod by hand or with pliers.

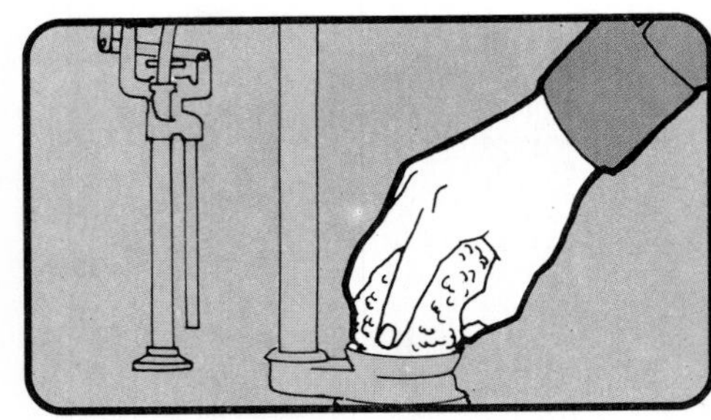

With the flush ball removed, clean the ball seat. Use fine steel wool for this, and buff the metal seat until it's shiny.

Adjust the linkage rod so it allows the ball to seat. To adjust the guide arm, loosen the setscrew as shown.

Align a flapper-type ball unit over the ball seat by twisting it on the overflow tube. A chain serves as the linkage wire.

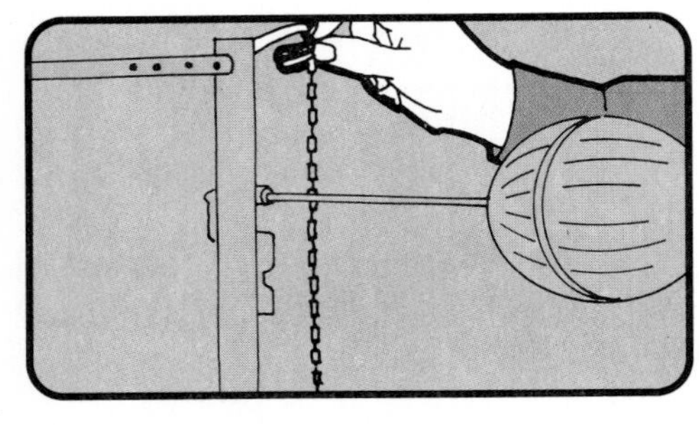

Adjust the lift chain to the proper length and fasten it to the trip arm. If the lift chain is broken or badly corroded, replace it.

Upgrading a Flushing Mechanism

Like anything mechanical, a flush-tank assembly eventually wears out. Its life-span depends on how often it's used and the hardness of the water in your area. Some water can quickly corrode parts, and lime deposits can quickly clog them up.

Almost all flush-tank mechanisms are replaceable by the piece, so you don't have to buy the entire unit. Flush balls, floats, lifts, and guides are "standard," so they usually fit any flush-tank make or model. However, if you're having trouble with the assembly, it might be smarter to replace the entire unit, a not-too-difficult project.

Several new flush-tank mechanisms depart from the traditional designs. One, a flapper-ball unit, features quietness of flush and an extremely long life-span. Another type—called a water-saver—doesn't use a float-ball component. Instead, water pressure regulates the water-inlet valve. This, in turn, meters out the exact amount of water needed in the tank for a full flush. This feature, in time, can save a considerable amount of water, and can eliminate the need to adjust a float arm and float ball.

If you choose to stick with a standard assembly, invest in a quality product. The cost may be a bit more at the outset, but your troubles with the unit should be minimal.

Adjusting Tank Linkage

No toilet flushing action can take place until you flip the handle on the tank. This mechanism is the key to the assembly, and it's also the most prone to malfunction. The tank linkage is made up of a handle, trip lever, linkage wire or chain, and connecting devices. If any one of them gives out, it affects the entire assembly.

Corrosion, the assembly's biggest enemy, usually occurs around the handle where it goes through the flush tank and connects to the trip lever. If you spot trouble here, remove the handle (a nut holds it tight) and clean the parts with fine steel wool. Then lightly coat the parts with a waterproof lubricant and reassemble them.

Be extremely careful when you remove the handle (use a wrench). Too much pressure can crack the flush tank. If this happens, you'll have to buy a whole new tank. If you can't remove the nut, you'll have to cut through the bolt with a hacksaw. Again, take care that you don't crack or chip the tank.

Trip-lever troubles start when the lever becomes bent or misaligned with the lift chain or linkage wire. The lever is set at a slight angle to the handle so it may operate freely without rubbing against the side of the tank, the inlet valve, or the overflow tube.

Flip the handle several times to make sure the trip lever is operating freely. If it isn't, try gently bending the arm toward the center of the tank for necessary clearance. As you bend the arm, hold it with one hand near the flushing handle.

The fastener or chain between the end of the trip lever and the linkage wire or chain often presents a problem. The water corrodes this part, and it actually rots away. Fortunately, replacing this connector is extremely simple. If the new part is made of brass, it will survive longer in the water.

Straighten the linkage wire connected to the trip lever. Lift the flush ball off the seat to prevent suction.

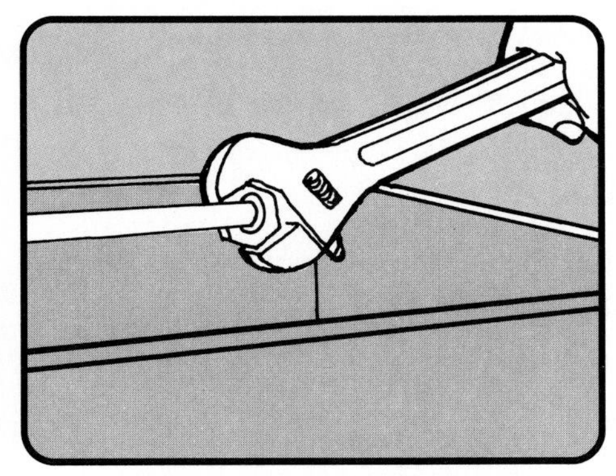

Take care when tightening or loosening the handle nut. Clean and coat the parts with waterproof lube.

Stopping Tank and Bowl Leaks

Occasionally a toilet will develop an external leak at one of three points—around the water supply pipe, where the tank joins the bowl, or around the base of the bowl.

To solve a water-supply-pipe problem, first tighten the nut that holds the fitting to the tank. If that doesn't work, shut off the water supply, remove the fittings, and install new washers, or a new pipe if necessary.

A leak where the tank joins the bowl may simply mean the tank's hold-down bolts have loosened. Drain the tank and try tightening the bolts as shown at right. If that doesn't stop the leak, you'll have to remove the bolts and install new washers.

With some older-model toilets, the tank is attached to the wall and connected to the bowl via an elbow fitting. If you spot leaks here, try tightening the elbow's slip nuts; if that doesn't work, repack them.

When a leak appears on the floor at the bowl's base, check the bowl's hold-down bolts. Chances are, they've loosened and allowed the bowl to rock on the seal underneath. Tightening the bolts might solve the problem; if not, you'll have to remove the entire toilet and install a new seal, as explained on page 45.

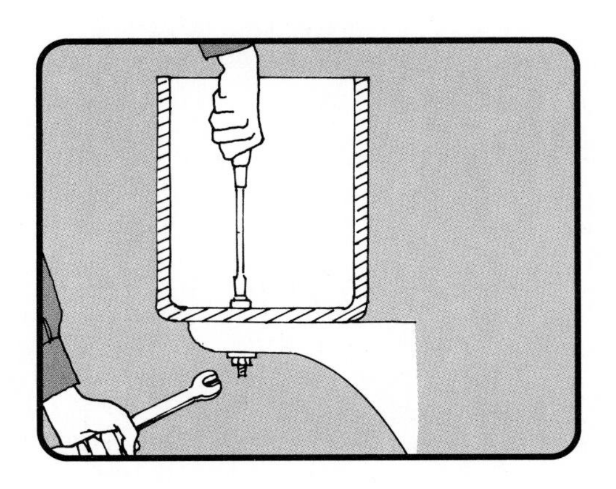

Apply penetrating oil to the tank bolts, loosen, then carefully retighten them. Too much pressure will crack the tank.

MAINTAINING A WATER HEATER

Today's water heaters generally provide years of trouble-free service, with or without maintenance. But just a little effort on your part can extend your water heater's life and cut down on its energy consumption.

Some manufacturers recommend that new gas or electric units be drained every two months for the first year they're in operation, then every six months after that. Doing this rids them of sediment, which builds up over time, impeding efficiency and providing you with less hot water at any given time.

To drain a water heater, first shut off the water supply by turning the shutoff valve at the top of the heater or at the meter. Next, place a bucket under the tank's drain valve, or fasten a garden hose to the valve and run the hose to a floor drain.

Now open the drain valve and let off water until it runs clear; then close the drain valve and open the supply valve.

You should also periodically check the heater's *pressure-relief valve* to be sure it's capable of letting off steam if pressure builds up in the tank. Just lift this valve's handle; if it's functioning properly, hot water will be released through the overflow.

If your heater is gas-fired, like the one shown here, inspect the flue assembly every six months or so. The draft divertor should be aligned with the flue, and the insulating tape that seals joints between flue sections should be intact.

The ports of a gas burner may have to be cleaned every two years or so—as explained in the chart below. And if you're plagued with a pilot light that just won't stay lit, adjust it according to the instructions that came with the heater.

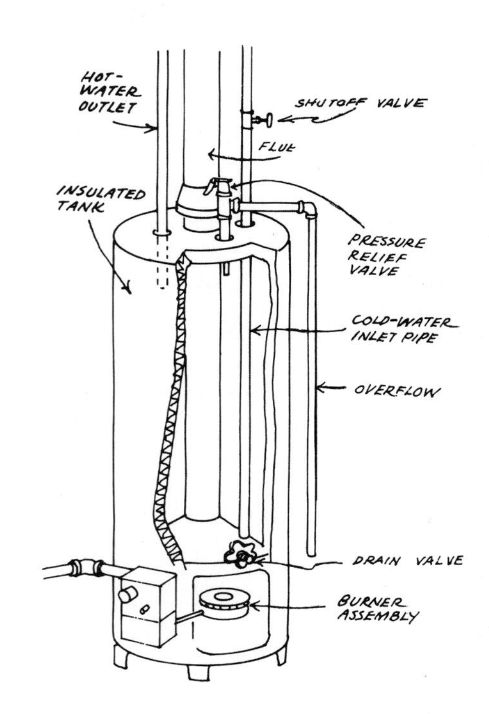

TROUBLESHOOTING WATER HEATERS

Problem	Solution
Water won't heat (electric)	Check the fuse box or circuit breaker for a blown fuse or a tripped switch; reactivate. If the heater continues to blow fuses or circuits often, call in a pro.
Water won't heat (gas)	Pilot light isn't burning; relight it. Unclog burner ports as explained below. Make sure the gas connection shutoff valve is fully open. Check temperature control knob for proper setting.
Water too hot	Check the thermostat setting; turn back setting, if necessary. The thermostat may be malfunctioning or not functioning. If you suspect this, call in a professional.
Water tank is leaking	Turn off the heater's water and gas or electrical supplies and drain the tank. It'll probably have to be replaced. To learn about installing a new unit yourself, see pages 46 and 47.
Water supply pipes leak	Tighten the pipe fittings. If this doesn't work, turn off the water and replace fittings. If water is condensing on the water supply pipe, wrap the pipe with standard pipe insulation.
Clogged gas burner ports	Remove the debris with a needle or the end of a paper clip. Do not use a wooden toothpick or peg; either can break off in the portholes.
Gas flame burns yellow	The burner may not be getting enough primary air. Also check the pilot light; the flame should be about ½ inch long. Call in a pro for any necessary adjustments. The burner of a gas water heater should be serviced professionally every 24 months or so.
Heater smells of gas	Immediately turn off the gas at the main supply valve. Open the windows and let the gas out. Turn on the gas at the main valve and coat the pipe connections with soapy water. If bubbles appear, the connection is leaking. Do not relight until the gas leak has been repaired.

WINTERIZING PLUMBING

Winter cold can wreak havoc with a plumbing system—and today most homes are plumbed with this in mind. Still, many homeowners have found out the hard way what happens to pipes in vacant houses with little or no heat.

Completely shutting down a plumbing system is neither difficult nor costly. Follow the procedures outlined here and in the drawing below.

On the day of the shutdown, turn off water at the meter or—better yet—schedule with the city water department to turn off service at the valve outside your home. (This is usually located in the front or backyard, and requires a special key or wrench to operate.)

Then, starting at the top of the water supply system, open every faucet—bathtub, shower, lavatories, and so on. Be sure you don't miss any you don't normally use, such as an outside sill cock (faucet) or underground sprinkler system. (These should be drained every fall anyway, whether your home will be heated or not.)

Turn off power to the water heater and drain it, too. By the time you reach your system's lowest point, its supply pipes should be completely empty. Make sure, though, that there's an outlet at the lowest point. This might be the water heater, a basement laundry tub or washing machine, or a valve installed specifically for draining the system.

Now you have to go through the entire house a second time to freeze-proof its drainage system. Start by removing the cleanout plugs on all sink and lavatory traps (see page 5), or, if a trap doesn't have a cleanout plug, dismantle and empty the trap itself, as shown on page 17. After you've emptied each trap, replace it or its cleanout plug, then pour in automotive antifreeze mixed with water in the proportions specified for cars in your climate.

You won't be able to drain some traps, such as the ones in toilets and perhaps those under tubs as well. With toilets, first flush them, pour a gallon of the antifreeze solution into each tank, then flush again. With bathtubs and other traps you can't get at, use the antifreeze full strength—at least a quart.

Water collects in dish- and clothes-washers, too. This you'll have to completely siphon out—but don't pour antifreeze into these appliances or any freshwater pipes. Finally, fill your home's main trap with antifreeze.

To refill a system that's been drained, you simply turn off all faucets, then open the water supply valve. Expect some sputtering at first as the water pushes air out of the lines. Don't worry about the traps; antifreeze in them will clear away automatically.

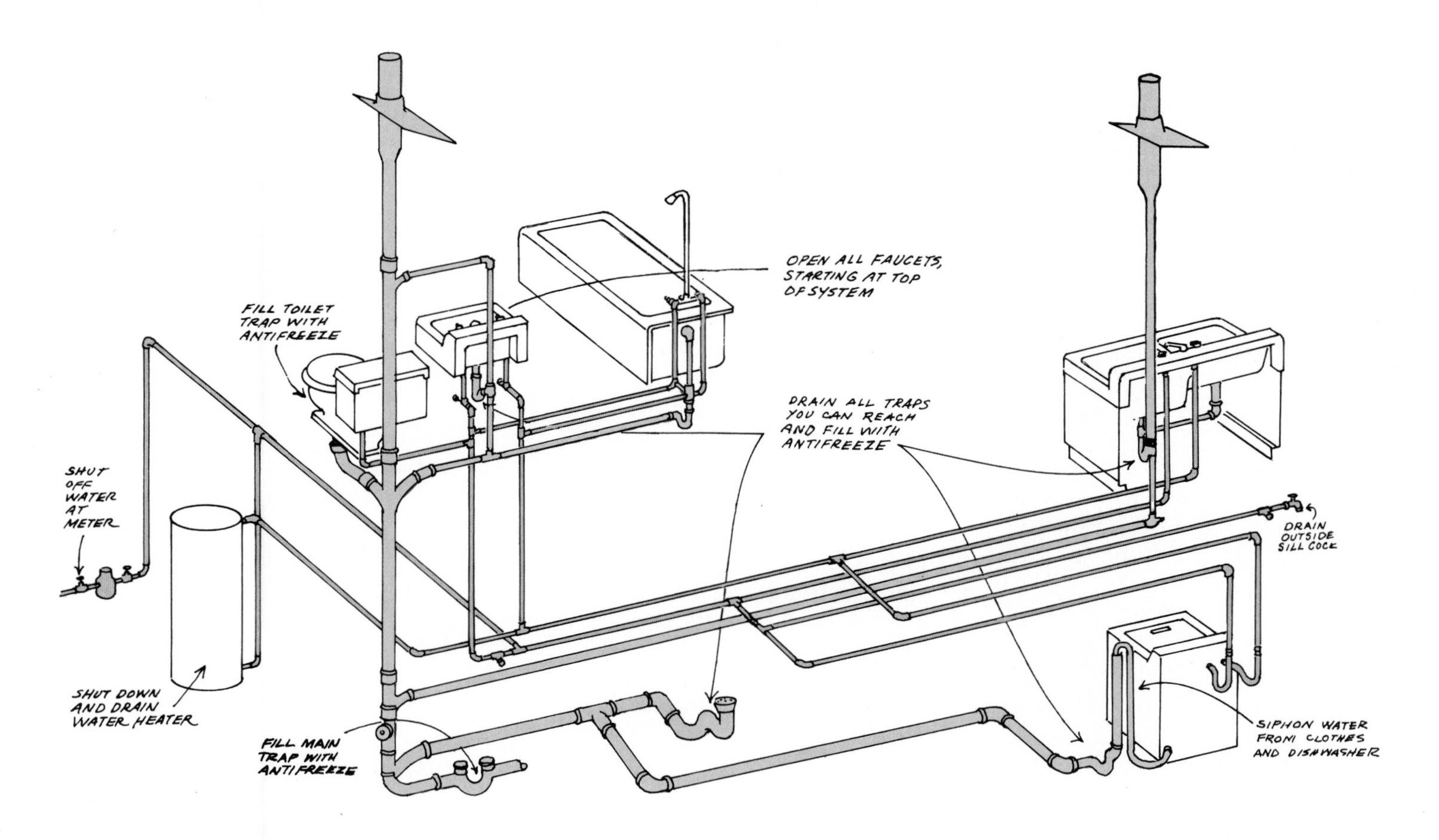

QUIETING NOISY PIPES

Pressure can get to any of us from time to time, and believe it or not, the same thing is true of your water system. Usually under the considerable load of 60 pounds of pressure per square inch (psi), your home's pipes can make a nerve-racking array of noises. Discussed below are the more common maladies and how you can cure them.

Water hammer, the loud bang you hear when you open a faucet, run the water, and quickly close the faucet, is terribly common. Automatic washing machines also produce this sound when a solenoid valve snaps shut. To prevent water hammer, most house fixtures have an air chamber. This can fill with water and banging begins. To fix, first drain the system. Then refill the pipes (the air chamber will fill with air again and shouldn't act up for several years).

If your system isn't outfitted with chambers, install one at the faucet fixture. This chamber provides a "cushion" of air on which the bang can bounce (see below).

Machine-gun rattle signals a faucet problem. Try replacing the washer.

A *whistle* indicates that a water valve somewhere in the system is partly closed. The water, under pressure, narrows at the valve and causes the whistle. Simply open the valve as far as you can. If a toilet whistles, adjust the inlet valve (see page 19).

If you hear *running water*, check for leaks at toilets, sill cocks (outdoor faucets), your furnace humidifier, and your water softener.

Generally, you can trace *soft ticking or cracking* to a hot water pipe that was cool, then suddenly was reheated with water. Muffle with insulation.

Bangs may result from water pressure in the pipes that causes the pipes to bang against their metal hangers. Have someone quickly open then close the faucet to cause a bang; often you can see the pipes move. To fix, see the suggestion below.

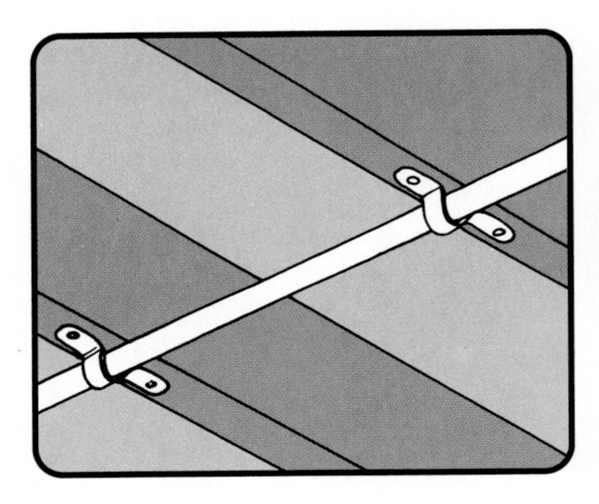

To stop bangs and squeaks, nail pipe hangers as shown. Be sure not to use galvanized hangers on copper.

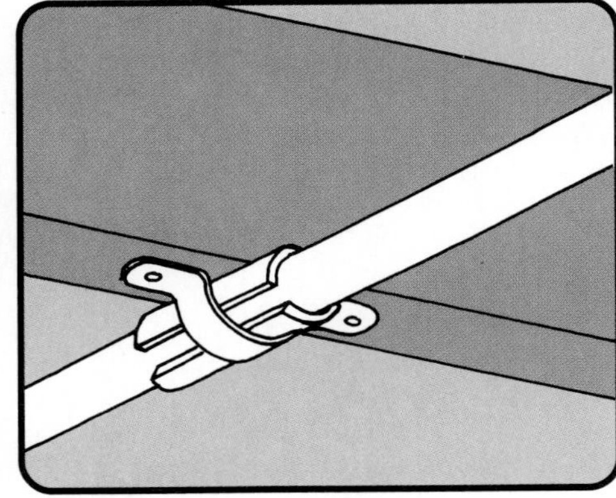

"Soundproof" pipes that touch hangers with short lengths of rubber hose. Split the hose lengthwise and slip it around the pipes.

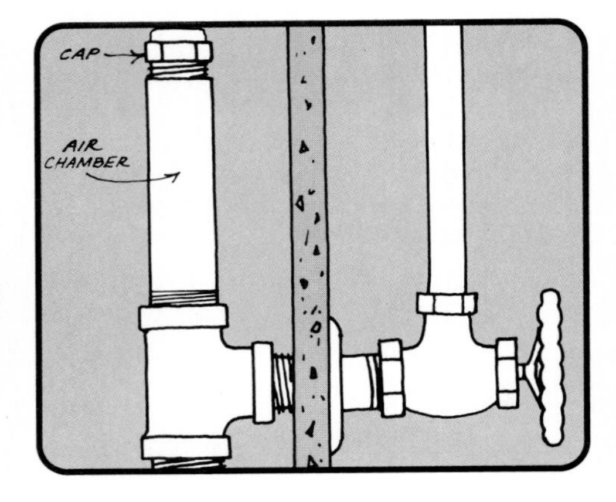

An air chamber is a length of pipe rising above the supply pipe, usually located near a faucet or fixture shutoff.

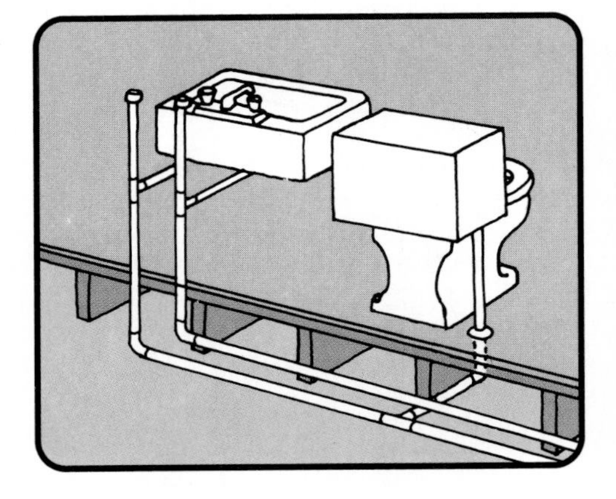

If you're extensively remodeling or building a new house, here's an air-chamber hookup diagram for a lavatory and toilet.

Copper-coil air chambers are available at plumbing stores. To install them, break the supply line and add a tee fitting.

COPING WITH PRESSURE PROBLEMS

Water pressure is one of those things you can have too much *or* too little of. Too little pressure results in trickles rather than streams of water, the results of which are obvious. Too much pressure, although a much rarer problem, can wreck faucets and weaken connections in your system.

If you have too much pressure in your lines, you need a pressure reducer, a device you can easily install.

If you're suffering from too little pressure, your immediate task is to locate the source of the problem. Start by removing aerators and shower heads from fixtures. If the strainers in these units are blocked with sediment and lime deposits, clean the strainers.

Next, make sure all shutoff valves are fully opened; partly closed valves slow water flow considerably.

If your water source is a well and you have an automatic pump, the pressure regulator at the pump may be set too low. Also check for a loose pump belt. In the winter months, low pressure can be caused by a frozen pipe or pressure switch.

If none of the obvious checks produces any results, break a connection in the water system. If you find lime deposits inside the pipe, you may have to finance a new plumbing job. Under no circumstances should you try to flush a limed system with chemicals. You can, however, have sediment flushed from the pipes, which may restore much of the pressure. This is a job for a professional plumber; don't attempt to do it yourself.

If liming is a problem in your area, the cheapest and easiest way to correct it is to install a water softener on cold as well as hot lines.

But before calling in a plumber, call water department officials and ask them to check the water main leading into your home. It could be faulty.

Make sure all supply valves are open. Also turn off water at the main, and check valve parts for damage, corrosion, or liming.

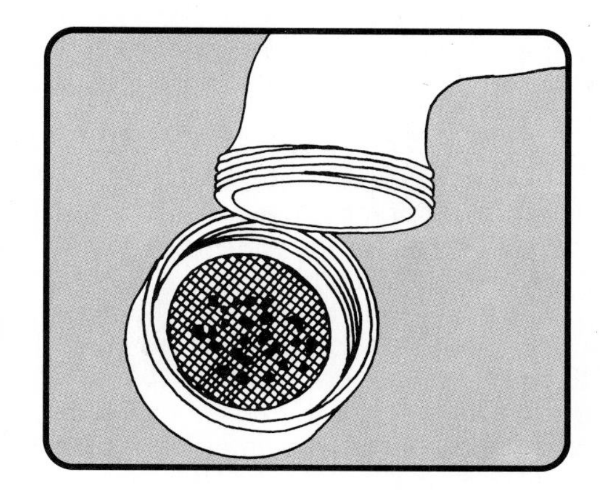

Unscrew aerators on faucet spouts to clean out any debris. If the wire strainer is badly corroded, replace it.

An inexpensive pressure gauge is a fast, easy way to check the water pressure. It should register 50 to 60 pounds per square inch.

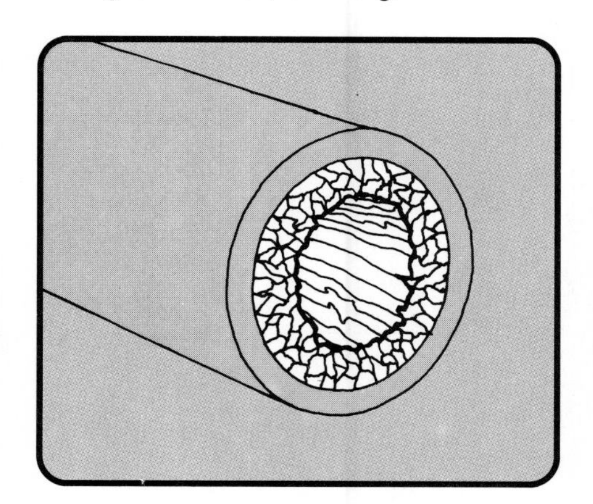

Limed pipes slow water to a trickle. If flushing the system doesn't help, you'll have to have your house re-plumbed.

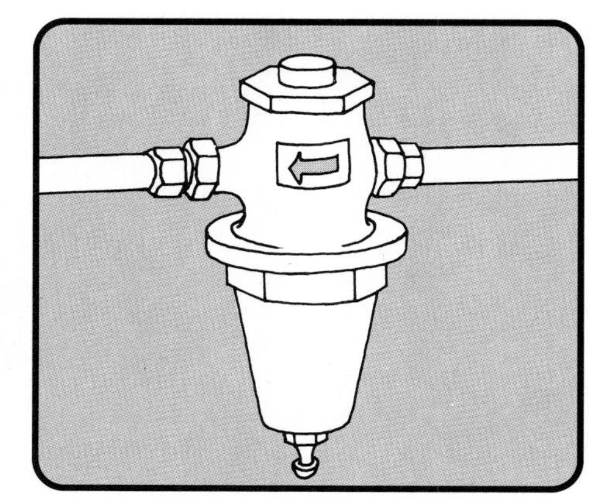

To decrease pressure, buy a pressure-reducing valve. The further open the valve, the less pressure you should have.

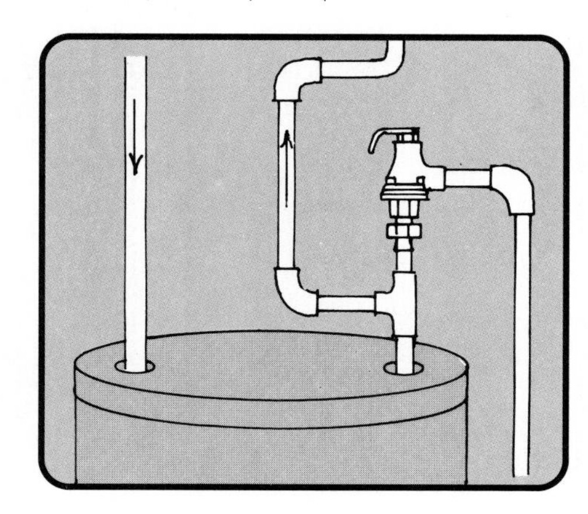

If your water heater doesn't have a relief valve, install one this way. Without one, hot water pressure could reach dangerous levels.

MAKING PLUMBING IMPROVEMENTS

Like carpenters, plumbers divide their work into two general categories—*roughing-in* and *finishing*. In the roughing-in stage, you cut sections of pipe to length and piece them together with a variety of standardized *fittings*. Then you finish off the job by hooking a *fixture* to the new lines.

The balance of this book takes you step by step through both phases of a plumbing project—from those critical first measurements to the moment you turn on the water and check your work for leaks.

The key to visualizing the way any plumbing run will go together lies with the fittings illustrated on the opposite page. Plumbing components are made from a wide variety of materials (see page 28), but all join with similar elbows, couplings, tees, and other connecting devices.

Note, though, that we've divided the chart into different sections for supply and DWV (drain-waste-vent) piping. These fittings are not interchangeable, even when they're made of the same materials. That's because drainage fittings have smooth insides. Since supply fittings are under pressure, their slight restrictions don't critically impede the flow.

MEASURING PIPES AND FITTINGS

Before you can buy the parts for any pipe-fitting project, you first have to know the diameters you'll be dealing with, and sometimes—especially if you're purchasing pre-threaded stock—the exact lengths as well. Computing both can be tricky until you get the hang of it.

Start by realizing that pipes are always sized according to their *inside* diameters. This means that the best way to get an accurate fix on what you need is to break open the run you'll be tying into and measure the pipe—not its fittings—as illustrated below.

Secondly, don't be surprised to discover that the inside diameter turns out to be slightly larger or smaller than a standard pipe size. So-called "one-inch" steel pipe, for instance, may be slightly greater or slightly less than an inch inside, depending on the thickness of its walls. Rounding off your measurement to the nearest ⅛ inch gives the *nominal dimension* you'll need.

The thing to keep in mind when you're figuring lengths is that you have to account for the distance each pipe engages in its fittings, as well as the distance between fittings. To do this, first measure from *face* to *face,* then add on the *socket* depths, as shown here. Socket depths vary somewhat from one pipe material to another, but remain the same for all fittings of a given material.

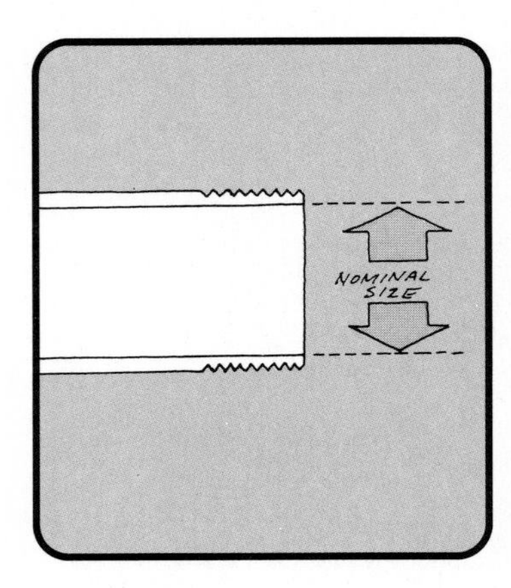

Always measure inside—not outside—pipe diameters. Actual and nominal dimensions can vary by 1/16 inch or so.

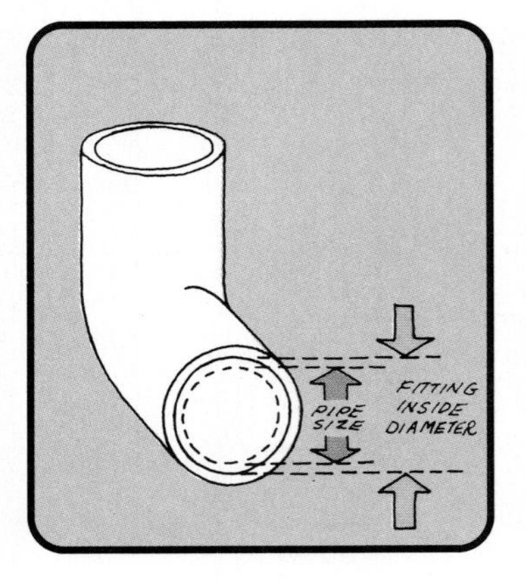

Fittings are also sized according to the pipe's inside dimensions, so measuring a fitting doesn't tell you much.

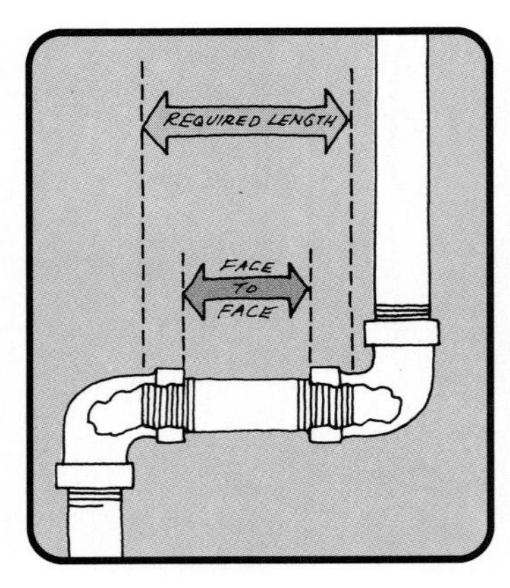

A pipe that's a little too short may leak at one or both ends. For accuracy, compute the length from face to face.

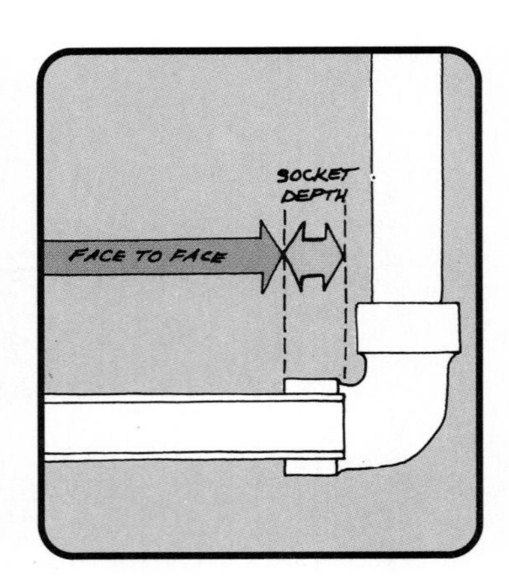

Now check the socket's depth. Since pipes have fittings on both ends, multiply by two, then add the face-to-face length.

KNOWING PLUMBING FITTINGS

Water Supply Fittings	Uses
Elbow	You'll need an elbow anywhere a plumbing run changes direction. Most make 90- or 45-degree turns. *Reducing elbows* (not shown) connect pipes of different diameters. *Street ells* have female connections on one end, male on the other. These let you couple one with another fitting.
Tee	Use this one wherever two pipes intersect. *Reducing tees* let you connect pipes of different diameters, as you might in taking a ½-inch branch off a ¾-inch main supply line, for example. To order, give the dimension of the main line first, then the branch line—¾x½ in the case cited here.
Coupling	This fitting connects lengths of pipe in a straight run. Once you've assembled pipe with couplings, you can't break into it without cutting at some point. *Reducers* let you step down from one pipe diameter to a smaller one. *Slip couplings* or *rings* (see page 37) let you connect to an existing copper or plastic line.
Union	Used mainly with threaded stock, a union compensates for the fact that all pipes have right-hand threads. You'll need at least one in any run of threaded pipe, and might want to add others so you can easily dismantle sections at a future time. More about this important fitting on page 37.
Cap, Plug, Bushing	To close off the end of a pipe, install a *cap;* these are available for both threaded and non-threaded pipe. To seal an opening in a threaded fitting, screw in a *plug*. Need to insert a pipe into a larger-diameter fitting? If so, a *bushing* is the answer. It's threaded both inside and out.
Nipple	Actually just lengths of threaded pipe that are less than 12 inches long, nipples are sold in standard sizes because short pieces are difficult to cut and thread. Get an assortment of these to join fittings that will be close together. A *close nipple* has threads from one end to the other for really tight situations.
Valve	*Gate valves* slide a gate-like disk that completely opens or closes the flow; these must be turned fully on or fully off. *Globe valves* operate like a compression-type faucet (see page 11); with these, you can regulate the flow as well as stop it. Even when wide open, globe valves constrict the flow somewhat.
DWV Fittings	**Uses**
Bend	Use these to change direction when running DWV pipe. Note that they have a gentle curve, rather than an abrupt angle where waste might lodge. Select ¼ bends for 90-degree turns, 1/5 for 72-degree angles, 1/6 for 60 degrees, ⅛ for 45, and 1/16 for 22.5. A *closet bend* connects a toilet to a main drain.
Tee, Wye, Cross	Available in a wide variety of shapes for different situations, these *sanitary branches* serve as the intersection where two or three drains converge. For vent piping, you can simply invert the same fittings. Like bends, these are shaped for a smooth downward flow of liquids and waste.

SELECTING PIPES AND FITTINGS

If you've been under the impression that pipe is pipe, prepare for a surprise before you check out the table below. It compares a dozen different materials.

Which type you'll want to purchase for a particular project depends first of all upon the plumbing code in your area. Begin by learning which materials are allowed and which ones aren't.

Next, consider what function the pipe and fittings must serve. Some can be used only in drain-waste-vent systems; others can't carry hot or drinking water. A few work for almost anything.

Finally, determine what the existing pipes in your home are made of. You needn't stick with the same thing—but you'll have to order special fittings to join dissimilar materials.

To help clarify matters, we've divided the dozen possibilities into four broad categories—copper, threaded, plastic, and cast iron.

Of the four, copper easily wins out as the most widely used, and virtually all codes permit it. Though somewhat more expensive than other types, it's lightweight, extremely versatile, and highly resistant to corrosion.

Threaded pipe—especially galvanized steel—has all but seen its day. Lifting, cutting, threading, and turning threaded pipe calls for lots of muscle work. What's more, steel limes up badly and can rust out in a decade or two. If you have an older home, chances are it was originally plumbed with galvanized steel.

If codes permit you to use plastic pipe, count yourself fortunate. Plastic is the easiest for amateurs to work with. You can't use it in every situation, though.

Cast iron pipe does only drain-waste-vent duty. It's by far the heaviest and most difficult to work with, though no-hub clamps make smaller jobs feasible for do-it-yourselfers.

COMPARING PIPE MATERIALS

Material		Uses	Joining Techniques/Features
Copper	Rigid	Hot and cold water lines; DWV	Usually sweat-soldered together, as illustrated on the opposite page. Light weight makes copper easy to handle.
	Flexible	Hot and cold water lines	Sweat-solder, flare, or use compression fittings (see page 30). Comes in long coils that can be easily bent. Too soft for exposed locations; fittings are relatively expensive.
Threaded	Galvanized steel	Hot and cold water lines; DWV. *Don't use as a gas line.*	Comes in standard 21-foot lengths that you cut, thread, and join with standard fittings. Cumbersome and time-consuming.
	Black steel	Gas and steam or hot water heating lines	Same joining techniques and features as galvanized, but as black pipe rusts readily, it's not used for household water.
	Brass and bronze	Hot and cold water lines	Again, you cut and thread. Because these are costly, they're rarely used in homes. Very high resistance to corrosion.
Plastic	Rigid ABS	DWV only	Cut with an ordinary saw, then solvent-weld sections together as shown on page 32. Lightweight and very easy to work with, but not all plumbing codes permit it.
	Rigid PVC	Cold water and DWV only	Same as ABS
	Rigid CPVC	Hot and cold water lines	Same as ABS
	Flexible polybutylene	Hot and cold water lines	Goes together with special fittings like the ones shown on page 33. Costly and not widely used in home plumbing systems.
	Flexible polyethylene	Cold water lines only. Used mainly for sprinkler systems.	Same as polybutylene
Cast Iron	Bell and spigot	DWV only	Joints are packed with oakum, then sealed with molten lead—a job for a professional plumber. More about this on page 34.
	No-hub	DWV only	You join sections with gaskets and clamps, as shown on page 35. Not overly difficult for an amateur to work with.

WORKING WITH RIGID COPPER PIPE

Arm yourself with a tubing cutter and a propane torch, master the knack of sweating joints illustrated below, and soon you'll be "running copper" like a professional plumber.

Rigid copper pipe comes in 10- and 20-foot lengths, in one of three wall thicknesses. The thinnest of these, Type M, is approved for interior use by most local codes, though a few of them call for medium-wall Type L. Type K has a thick wall and is required for applications that run underground. Fittings include almost all of those that are shown on page 27.

Most pros preassemble runs of copper pipe, fluxing and dry-fitting each joint, but not soldering yet. When they're satisfied that everything fits properly, they go back and torch it all together.

After you sweat each joint, carefully look for any gaps around its perimeter. Leaks that don't show up until after you turn on the water mean you have to completely drain and dry out the joint before you can re-sweat it.

When you're using the torch, protect flammable surfaces with sheet metal, asbestos board, or wet rags.

A tubing cutter does a fast, neat job of cutting pipe to length. If you use a hacksaw, take care to keep the cut square.

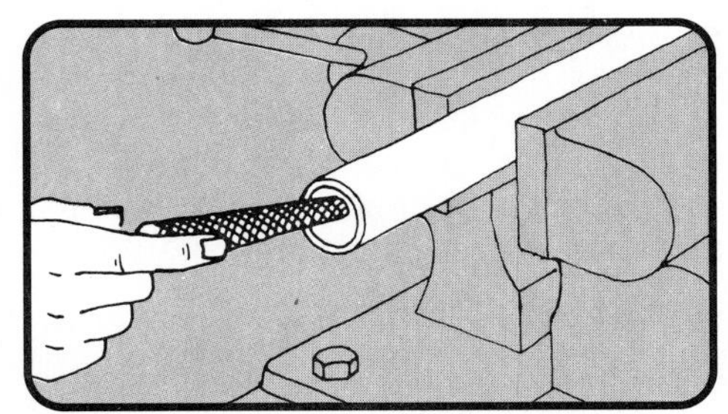

Remove all metal burrs from the cut with a file. Don't nick the metal; it could cause the connection to leak.

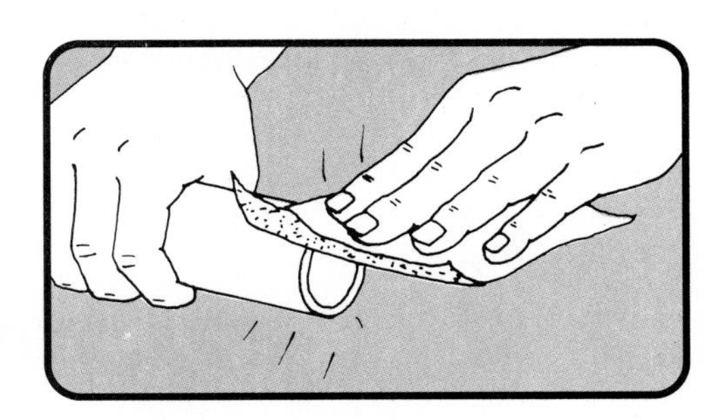

Shine the ends of the pipe to be connected with a fine-grit abrasive or steel wool. This will remove grease and dirt.

Apply rosin- (not acid-) flux soldering paste to the outside of the pipe and to the inside of the pipe fitting.

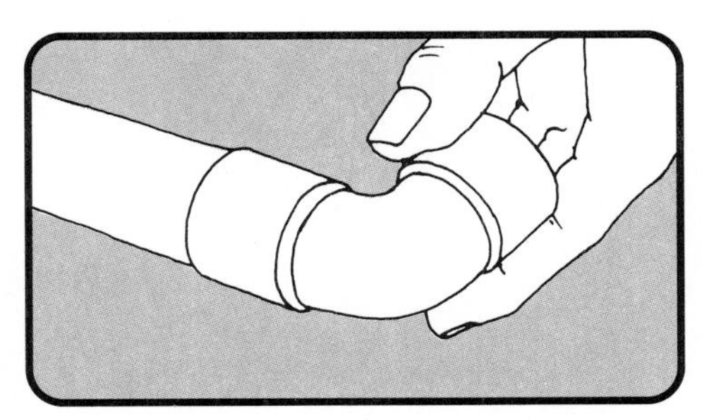

Slip the fitting onto the pipe. If the fitting has a hub or shoulder, make sure the pipe is seated against it.

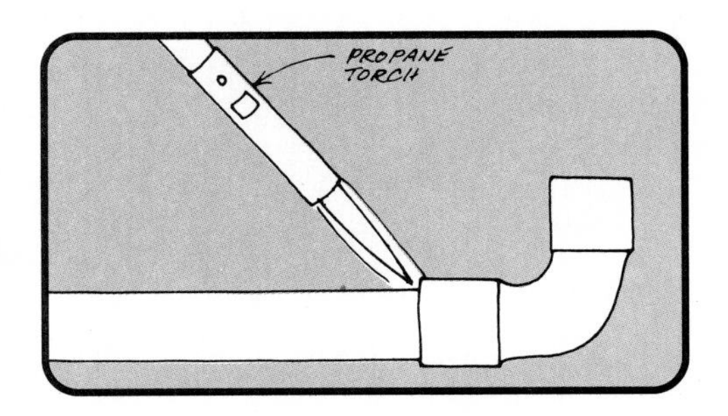

Heat the pipe and fitting where both components join. The tip of the inner flame produces the most heat. Don't overheat.

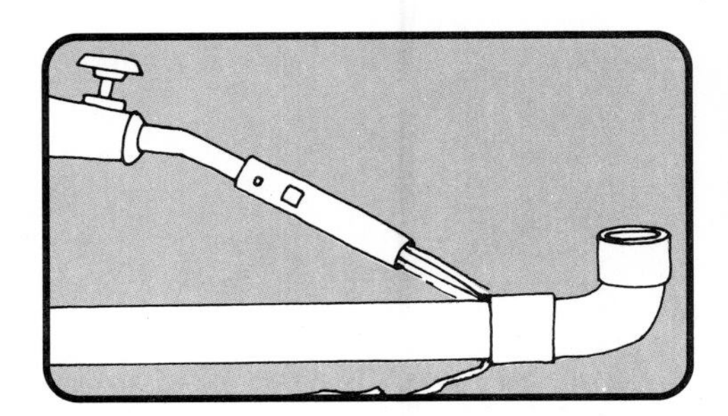

Test the pipe for temperature. If the solder melts, the temperature is right. The solder will flow into the connection.

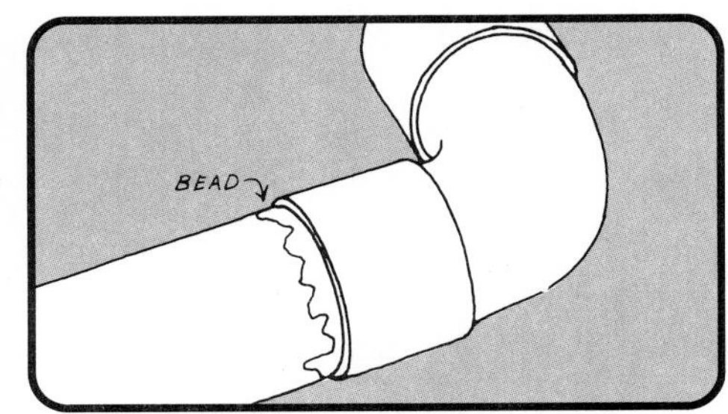

The solder will form a solid bead around the connection. When the bead is complete, remove the solder and heat from the pipe.

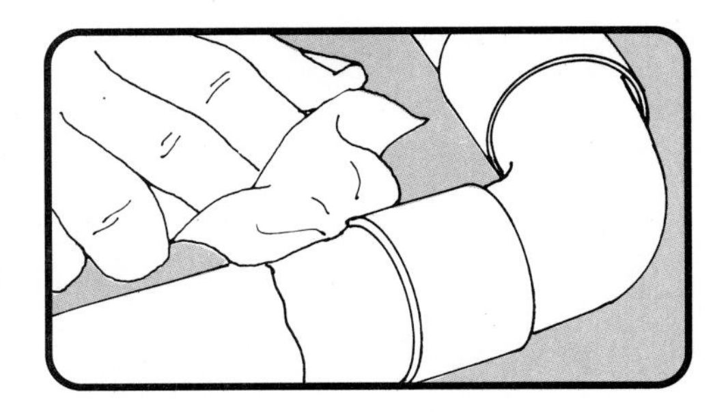

To give the joint a professional appearance, wipe it with a cloth as shown. Be careful not to get burned, though.

WORKING WITH FLEXIBLE COPPER TUBING

As its name implies, flexible copper tubing distinguishes itself by being exceptionally workable. It goes places that rigid copper pipe just can't, and so is ideal for many close-quarters situations. Do use care when bending it, though. Flexible copper tubing kinks easily, and once kinked it's practically impossible to bend back into its original round shape. Don't use kinked tubing—it impedes water flow.

There are two weights of flexible copper tubing: Type K, generally used for underground installations, and Type L, the interior product. Both generally come in 15-, 30-, and 60-foot rolls at plumbing and home center stores.

You can assemble runs with solder and standard connections (as shown on page 29), with flare fittings, or with compression fittings. Flare and compression fittings cost more than solder fittings, but they do assemble more easily and don't require heat, a big advantage when working in close quarters between wall studs and floor joists.

Consider soldering connections you'll never need to break; save the more costly fittings for hooking up fixtures that might need to be removed someday.

To form flare fittings, you'll need a flaring tool, a fairly inexpensive item (see below). Once you've flared the tubing ends, you just screw special fittings together. Be sure to put flare nuts on the tubing before flaring the ends, though.

With compression fittings, the pipe doesn't have to be flared. You simply slip a compression ring over the tubing. The ring compresses (and seals) when the fittings are screwed together.
The drawings below show how to join flexible tubing with either flare or compression fittings.

Bend flexible copper tubing over your knee, or use a spring tube bender. Work with gentle bends, not acute angles.

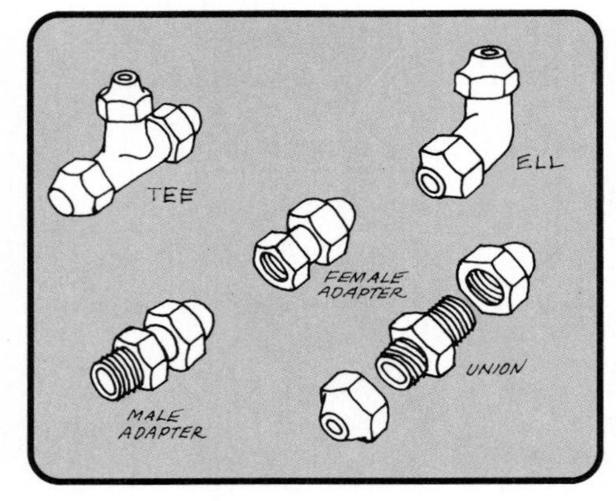

Fittings include tees, ells, unions, and a variety of copper, steel, and plastic adapters for hooking to other types of pipe.

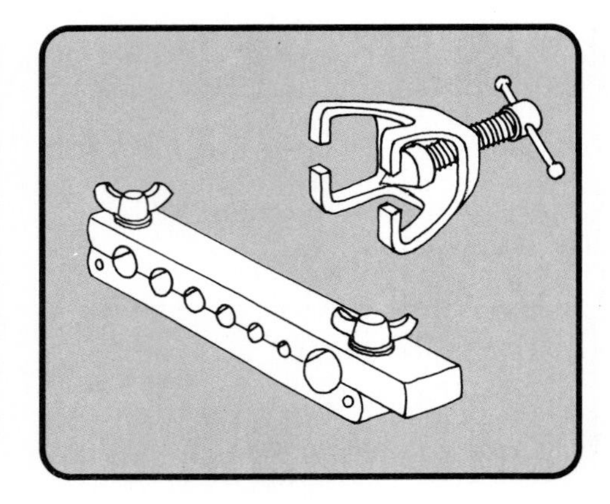

A flaring tool is a two-piece unit. Different-sized, beveled holes in the block accommodate several different sizes of tubing.

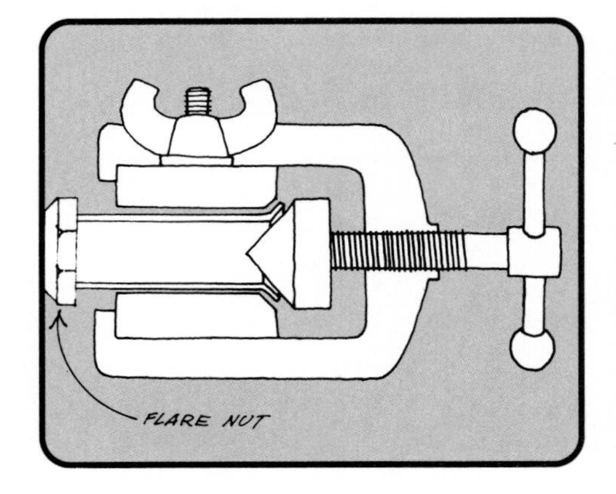

Clamp the tubing in the block. Attach and center the flaring part of the tool over the tubing. Then tighten, turning clockwise.

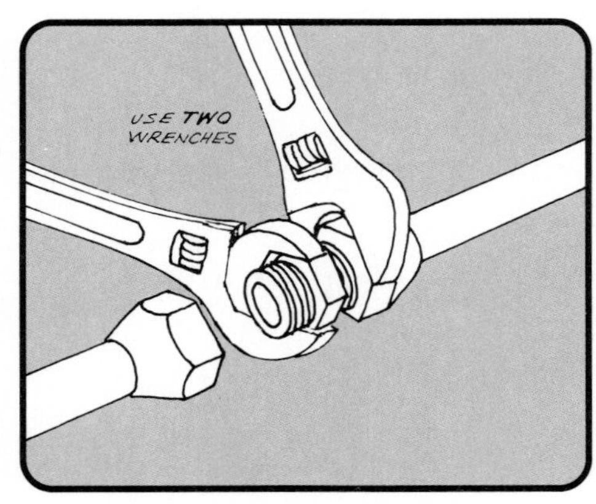

After the flares are made, use two wrenches to connect the joint. If the joint leaks under water pressure, re-flare the joint.

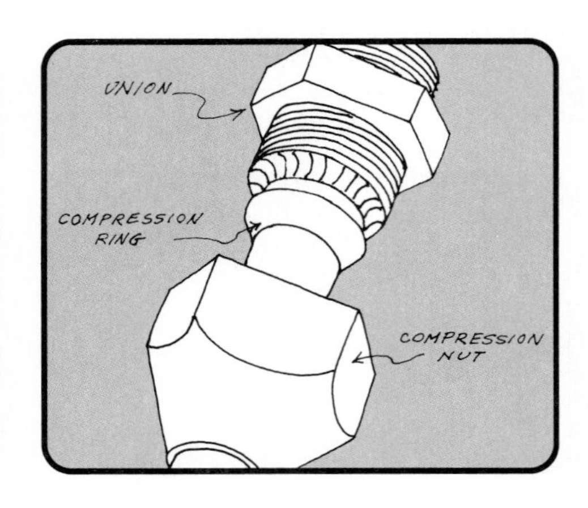

A compression joint has a ring that compresses under pressure of the fitting, producing a leak-free joint.

WORKING WITH THREADED PIPE

Though copper and plastic now offer better ways to go, plumbers have been threading together steel, iron, brass, and bronze pipes for a century or so—and unless yours is a fairly new home, at least some of its plumbing probably uses this old-fashioned system.

Need only a few sections of threaded pipe? Measure carefully, then order them precut and threaded. For bigger jobs, it may pay to thread your own.

Threading tools (these are best rented) include a vise, pipe cutter (you also can use a hacksaw), a threading die, die stock, reamer, and two 10-inch pipe wrenches.

The key to threading pipe is to start the die squarely on the pipe. You must get it right the first time or the threads won't thread, so take your time until you get the hang of it.

The size of the pipe you're using determines how far it must be turned into the fittings. For 1/8-inch pipe, the distance is 1/4 inch. For 1/4- and 3/8-inch pipe, it's 3/8 inch. One half- and 3/4-inch pipes need 1/2 inch of threads; and 1-inch pipe requires a 9/16-inch distance.

To measure the length of pipe needed, figure the distance between the face of each fitting, plus the distance into the fittings. For example, if you're working with 3/4-inch pipe and the distance between the face of each fitting is 4 feet, you'll need a piece of pipe 4 feet, 1 inch long.

If you're replacing a piece of pipe and threading the pipe yourself, remember to include a union fitting in the measurement. The same threading distance applies to unions as for the other fittings.

When threading brass pipe, you must be extra careful not to scratch or mar the finish as you thread it. And when cutting and threading galvanized or black pipe, be sure to wear gloves to protect your hands against sharp metal slivers and shavings.

To cut pipe, first clamp it tightly in a vise. Then make the cut with a pipe cutter. The sharp metal wheel does the cutting.

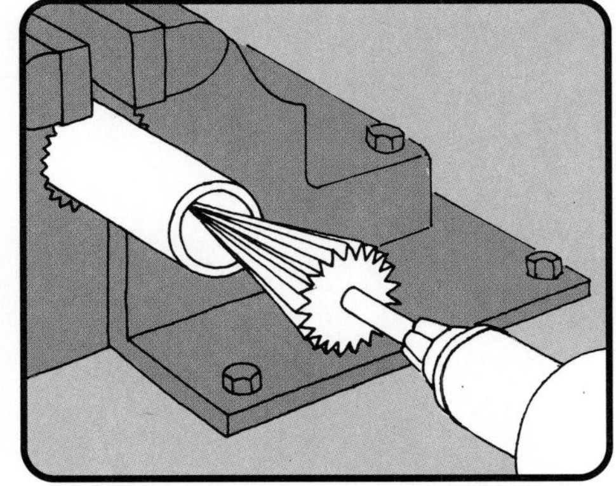

Remove burrs from the inside of the pipe with a reamer powered by a brace or drill. Remove just the burrs, not the metal.

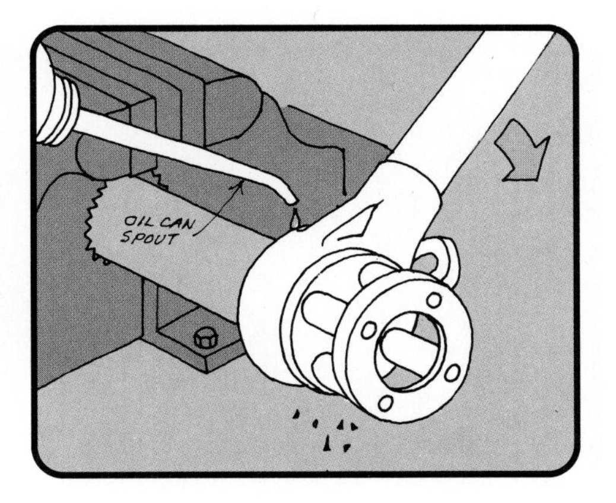

Lock the threading die in the die stock. Place the die squarely on the pipe and begin turning it. Apply oil as you cut the threads.

Apply pipe dope or pipe tape to the pipe threads (but not to the fitting threads) before twisting on the fitting.

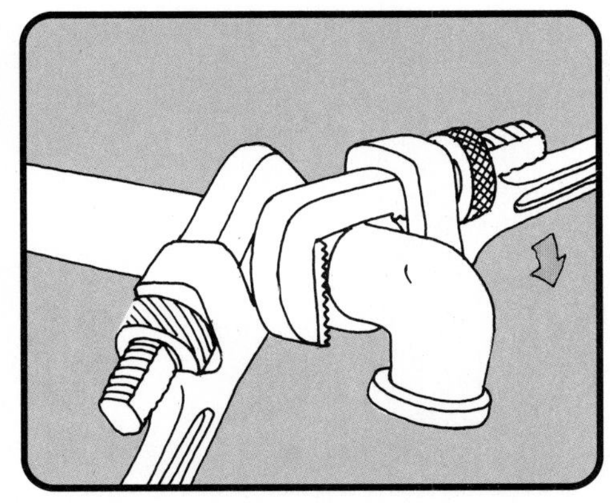

Using two pipe wrenches as shown above, turn the fitting onto the pipe. To avoid leaks, tighten down as far as you can.

WORKING WITH RIGID PLASTIC PIPE

Rigid plastic pipe, though only a recent introduction to the plumbing industry, offers great promise for pros and do-it-yourselfers alike. It's the lightest of all pipe materials, cuts with an ordinary saw, and glues together with fittings and a special solvent cement.

Be warned, though, that since plastic pipe hasn't yet been subjected to the test of time, many localities still prohibit or restrict its use—so before you spend a dime on plastic pipe or fittings, check the codes in your area to find out if they're allowed and in which situations.

And note that of the three types available—*CPVC* (Chlorinated Polyvinyl Chloride), *PVC* (Polyvinyl Chloride), and *ABS* (Acrylonitrile Butadiene Styrene)—only CPVC can be used for hot water supply lines. Don't mix these materials; they require different cements and expand at different rates.

Keep expansion in mind, too, when you're assembling runs of plastic pipe, especially CPVC. Provide plenty of clearance between fittings and framing, and bore oversize holes everywhere the pipes pass through wood. Otherwise, the system will creak, squeak, groan, and maybe even leak every time you turn on a water tap or run hot water into a drain line.

Check the drawings below and you'll see that solvent-welding plastic fittings calls for no special tools or skills. Alignment is critical, though, because once you've made a connection, the joint is permanent. Make a mistake and you have no choice but to cut out the fitting, throw it away, and install a brand-new one.

Realize, too, that solvent-welded joints don't reach working strength for periods of anywhere from 16 to 48 hours. This means your family may have to do without running water for a full day or even two after you've completed an installation.

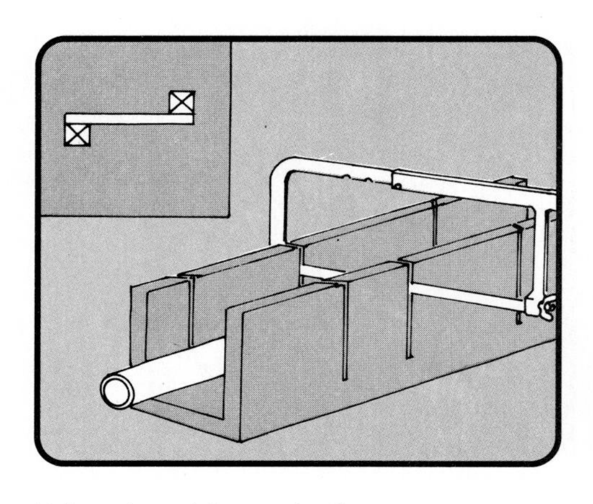

Using a bench jig or miter box, cut rigid plastic pipe with a hacksaw. Be sure the cut is absolutely square.

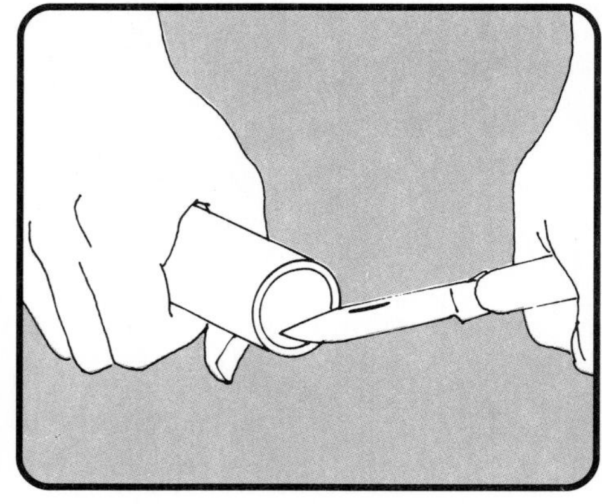

Very carefully remove the burrs made by the hacksaw with a sharp knife. Keep the pipe tipped downward so debris will fall out.

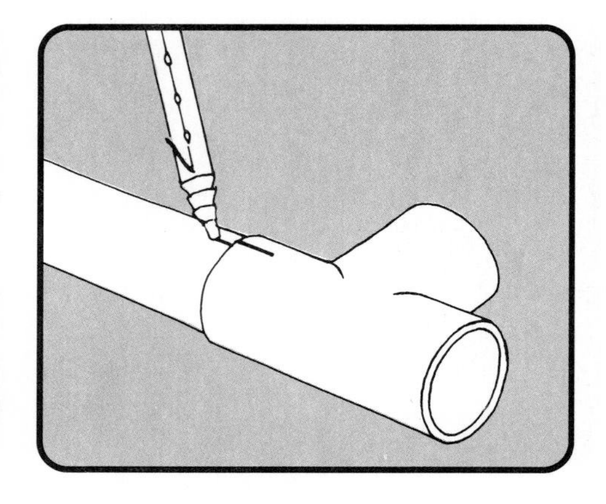

To ensure proper alignment of the pipe and fitting, dry-assemble them before applying the solvent. Mark the alignment as shown.

Very lightly sand the outside tip of the pipe. Then wipe it with a clean cloth and apply cement to both the pipe and connection.

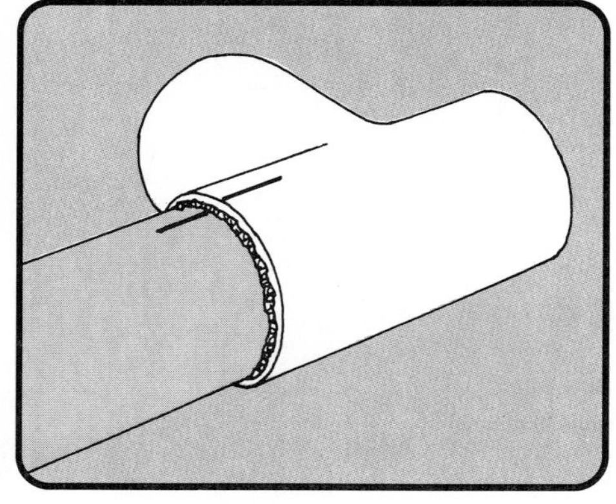

As you assemble the connection, give it a quarter turn to distribute the cement. A small bead should form around the fitting.

WORKING WITH FLEXIBLE PLASTIC PIPE

Of all the types of pipe available today, flexible plastic has seen less duty than any other. In fact, most people have never even heard of it. That's due to two factors. First, it's costly, and second, there are some limitations as to where it can be used. For example, only the pipe made from polybutylene can stand up to the heat generated by a hot water line. The polyethylene type won't.

If you do decide to use flexible plastic pipe, you'll find that it's available in three different water-pressure ratings: 125 psi; 100 psi; and a low-pressure product, which is so specified.

Unlike its rigid cousin, flexible plastic pipe goes together with clamps instead of cement. You can buy a potpourri of fittings—tees, ells, and straight connections—in a variety of diameters to fit various pipe sizes. You also can buy conversion fittings that let you join the pipe to copper and galvanized plumbing.

Although flexible plastic pipe is "flexible," don't overdo it; it can kink. Rather, bend it into gentle curves. The product is rigid enough to support itself on fairly long runs, although it's usually smart to use hangers for support.

There are no special tricks or stunts involved in working with flexible plastic pipe (see the basic techniques below). You should, however, keep in mind that the polyethylene pipe "softens" somewhat when it becomes warm. Connections on above-ground installations can expand in the hot summer months, causing leaks. Check these connections from time to time and tighten the holding clamps if needed.

Flexible plastic pipe that's exposed to the elements should be drained during cold weather to prevent frozen water from cracking and breaking it. But since the plastic is weather-resistant, you needn't disconnect the system and store it during cold weather.

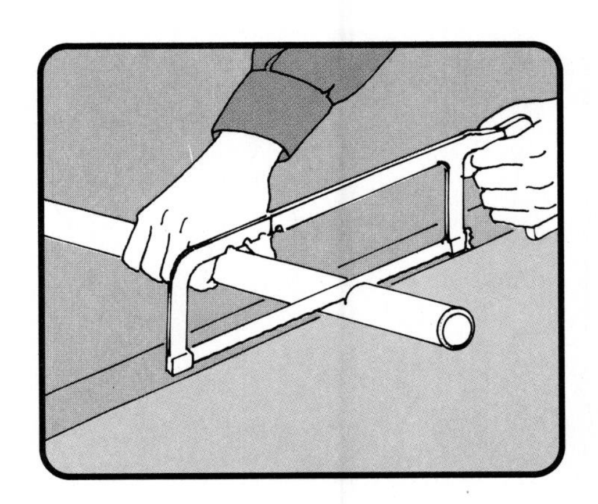

A hacksaw is the best tool for cutting flexible plastic pipe. Make the cut square, and remove saw burrs with a sharp knife.

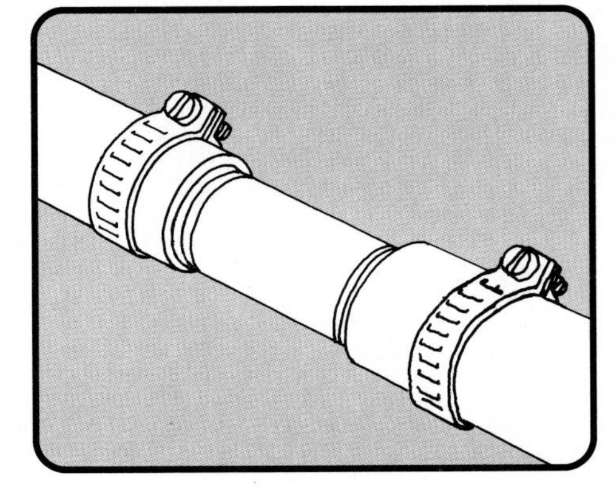

Use stainless steel worm-type clamps to fasten fittings to the pipe. Slip the clamp over the pipe before you add the fitting.

Position the clamp on the pipe so the clamp tightens against the entire shoulder of the fitting. This keeps the pressure even.

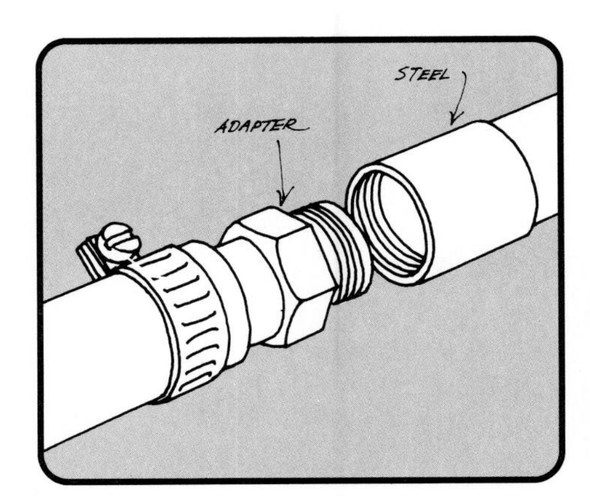

Always use an adapter when joining plastic to other materials. You'll need both pipe and adjustable wrenches for this job.

If you want to break a clamped connection, soak the connection in hot water to loosen it. Do not use a propane torch for this.

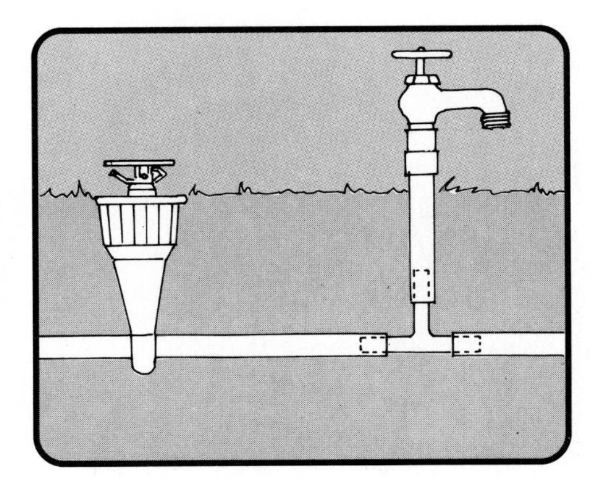

A hookup for an underground sprinkler utilizes tees and ells to turn corners. Tight bends will kink the pipe.

WORKING WITH CAST IRON PIPE

Until recently, installing cast iron DWV (drain-waste-vent) systems was all but out of the question for amateur plumbers. Traditional "hub-and-spigot" joints—a lipped *spigot* on one section slips into a bell-shaped *hub* on the other—had to be packed with oakum, then caulked with molten lead.

With one newer system, joints are sealed by inserting a neoprene gasket into the hub end of a pipe then forcing the bald (straight) end of the next pipe into it.

The easiest system, though, employs *no-hub* cast iron pipes. With these, you simply couple sections with a special neoprene gasket and an automotive-type clamp (see opposite page). You need no special tools, and if a fitting comes out a little cockeyed, it takes only a minute to loosen the clamp, twist everything the way you want it, and retighten it.

Best of all, no-hub components are compatible with hub-and-spigot piping, so if you ever have to cut into an existing cast iron soil pipe, you can simply cut away a section and slip in a no-hub tee (see the opposite page).

No-hub joints eliminate most, but not all, of the hassles involved in working with cast iron. Wrestling 5- and 10-foot lengths of this material into place is heavy work. Cutting can be a problem, too, unless you rent specialized equipment, or measure and buy precut lengths.

That's why, faced with a sizable DWV installation calling for a new soil stack, you might be wise to have a pro install the stack. Then you can run lateral drain and vent lines from fixtures to the stack.

Plan these runs so they'll require a minimum of bends, and be sure to pitch drain lines back to the stack. (For more about DWV systems, see pages 36–38.)

Be sure, too, that you properly brace the pipes with supports at each fitting and every four feet on straight runs. Drawings on the opposite page show commonly available hangers.

To cut cast iron, measure the length you need. Then scribe the cutoff line on the pipe with a wax-type pencil.

Elevate the pipe by placing it on a piece of 2x4 or 2x6. Make a 1/16-inch cut around the pipe's circumference.

Along the hacksaw kerf, rap the pipe with a baby sledge. Tap the pipe smartly; it should fracture cleanly at the cut.

A jagged break can be evened up by tapping with a cold chisel. You don't have to get the edges glassy smooth.

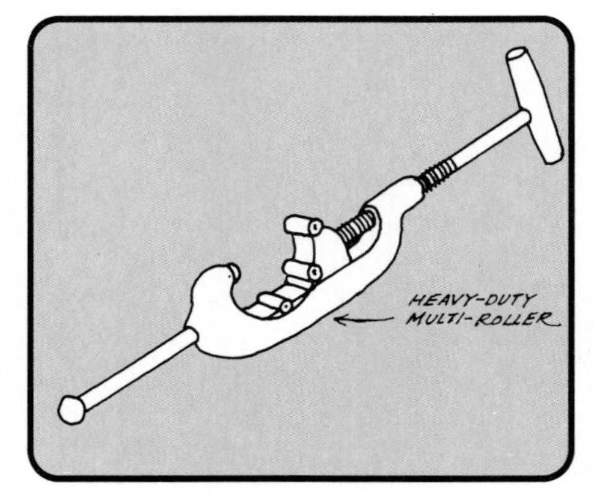

For small-diameter cast iron pipe, you can rent a roller-type cutter like this one. It will save you a lot of effort.

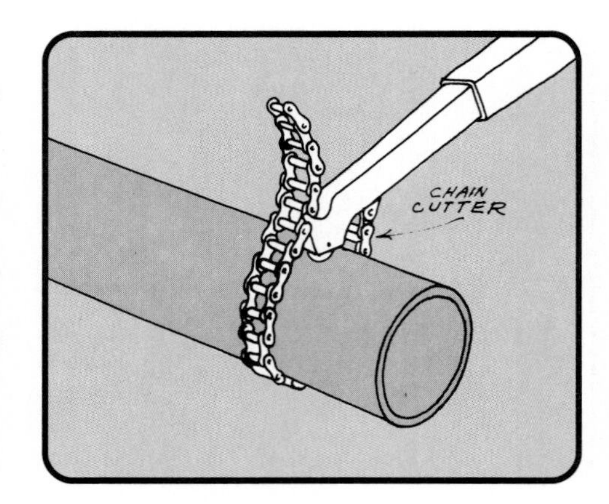

A chain-type cutter handles larger diameters. The chain secures the pipe while the cutting wheel scores the surface.

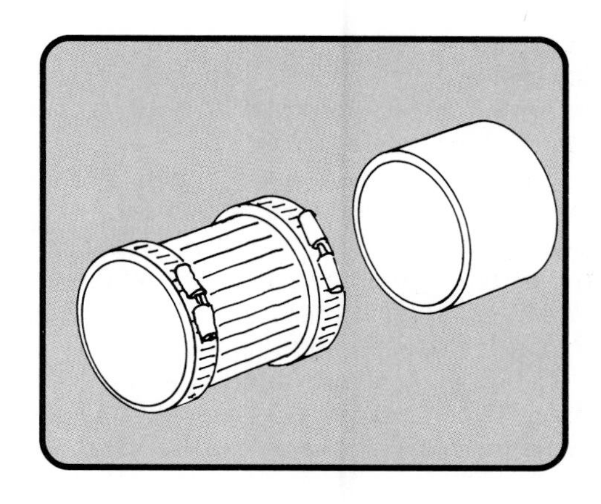

A no-hub connector consists of a neoprene sleeve and a metal clamp that tightens with two worm-drive screws.

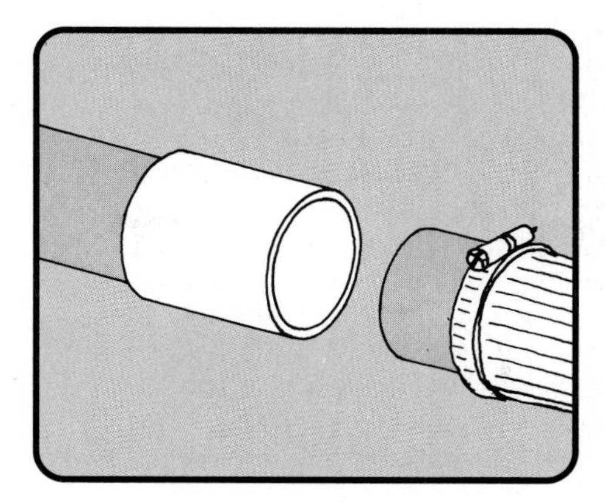

To assemble joints, remove any burrs from the pipes' cut ends, then slip the clamp and sleeve onto the pipes as shown.

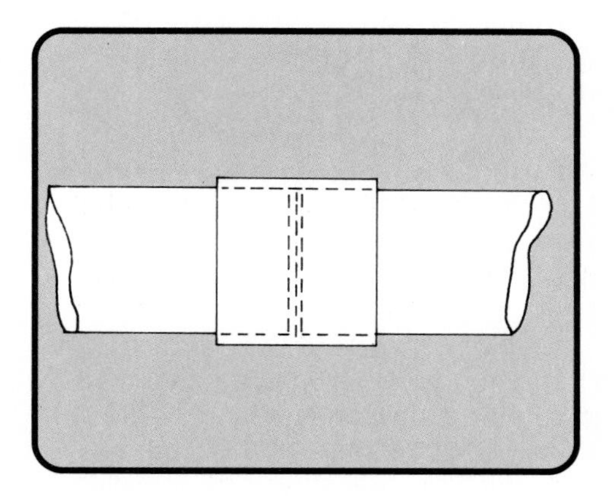

Be sure when you shove the pipes together that their cut ends butt together, separated only by the sleeve's inner ridge.

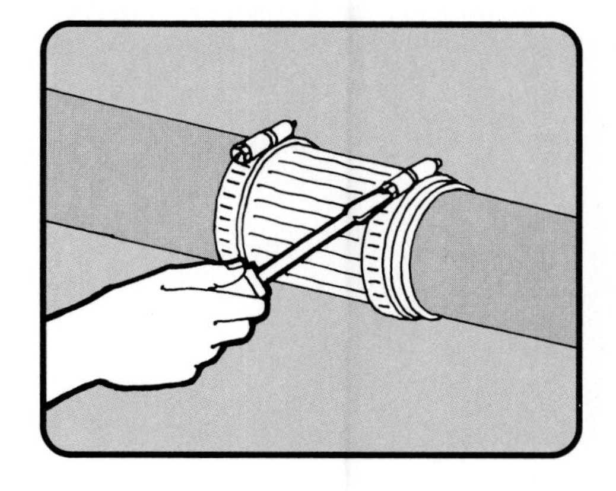

Now center the clamp and tighten it securely with a screwdriver or special T-wrench. Be careful not to overtighten, though.

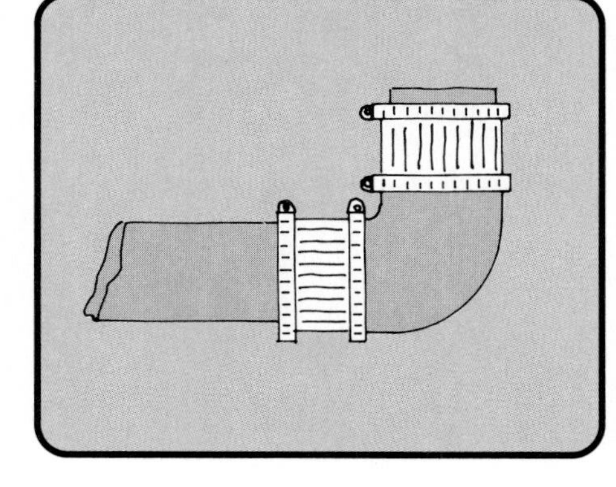

Bends and all other no-hub fittings go together the same way. You'll need a sleeve and clamp for each end of the fitting.

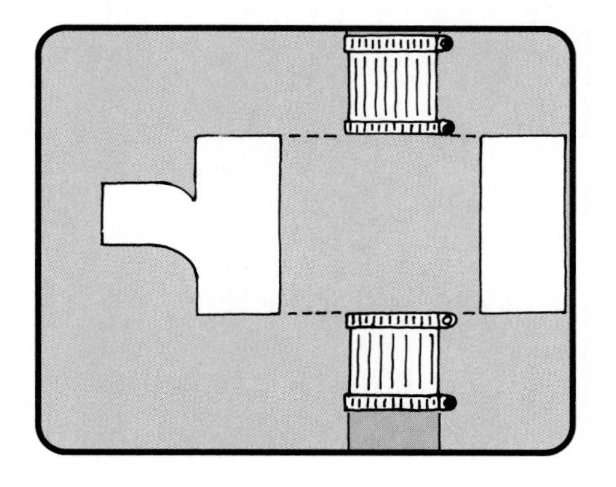

Here's how to fit a no-hub tee into an existing pipe run. Slip clamps over the cut ends, then position the fitting.

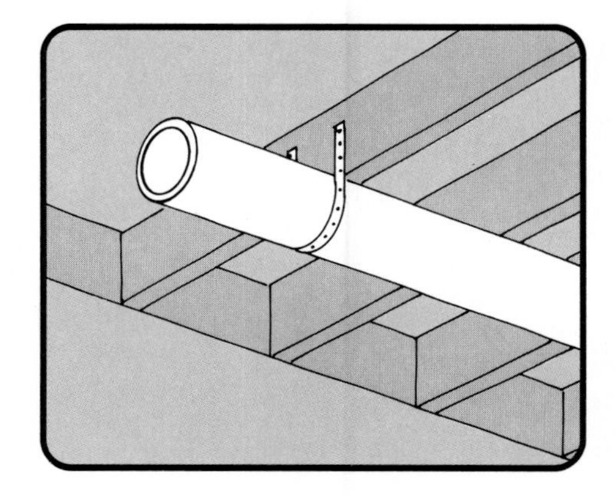

Support horizontal runs with pipe straps or hangers. This type attaches to the faces of floor joists.

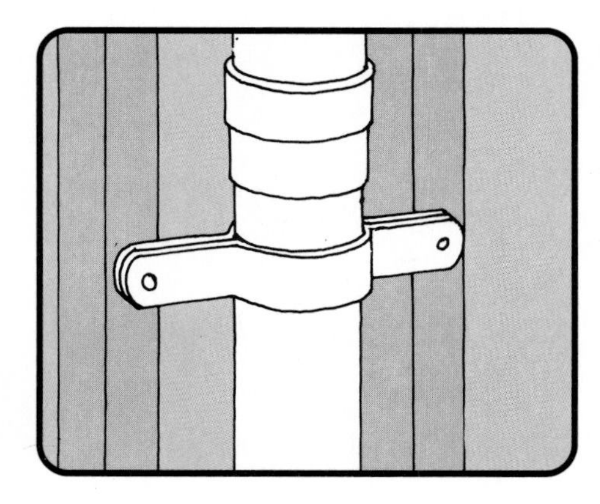

Two-piece hangers serve as a clamp to hold vertical runs. Install one below each joint and at floor level.

CONNECTING TO DWV LINES

Thinking about adding a new fixture at your house? If so, you've probably already wondered where the water's going to come from—and where it's going to go.

Of the two parts to this problem, the first is relatively easy to solve. Small-diameter supply lines can zigzag easily through tight spots—and because they're under pressure, you can extend them almost any distance.

Drain-waste-vent lines are another matter. First, they're much larger and more difficult to conceal. Second—and even more critical—you have to ensure that any new fixture is properly vented. Otherwise, vacuum in the drainage system could suck water from the fixture's trap and let sewer gas back up into the house.

Does this mean you have to install a vent, as well as a drain, for each and every fixture? Fortunately, most plumbing codes let you dispense with individual venting under certain circumstances (see below).

Note that the most common of these exceptions—*wet venting*—requires locating the fixture within a specified distance from your home's soil stack—and the fixture must drain only liquid wastes. This means you can wet-vent a lavatory, tub, or shower, but not a toilet or kitchen sink. Get advice from a pro before you plan DWV hookups for these solid-waste carriers.

Clearly, it makes sense to situate a new fixture as close as possible to existing lines. Your alternative is to opt for an entirely new stack—one you might consider if you're contemplating a major installation such as a new bath.

When you're measuring out distances, don't neglect to allow for a slope of ¼ inch per foot (or whatever local code calls for) from the fixture to the main drain. You may end up having to elevate a tub or shower to provide proper drainage.

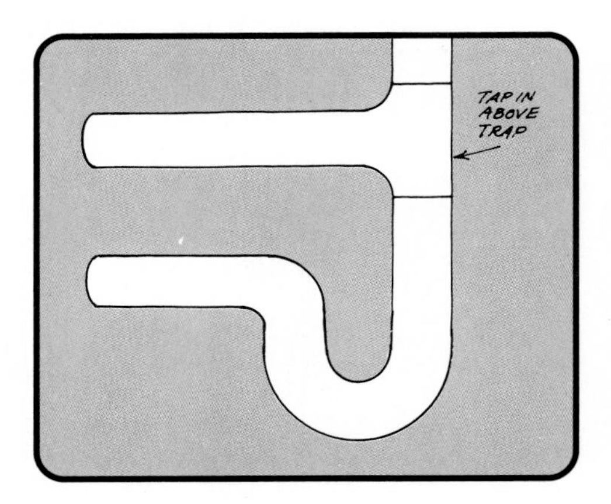

Often you can drain two lavatories or sinks into the same trap—provided their drain outlets are no more than 30 inches apart.

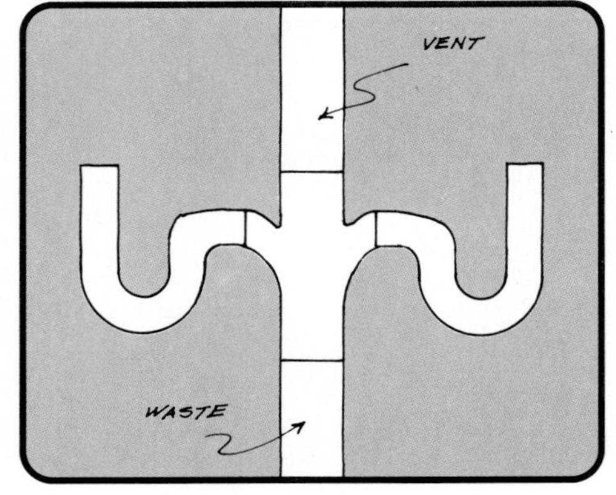

Unit venting lets you locate fixtures back to back, discharging into the same vent-waste line. Each has its own trap.

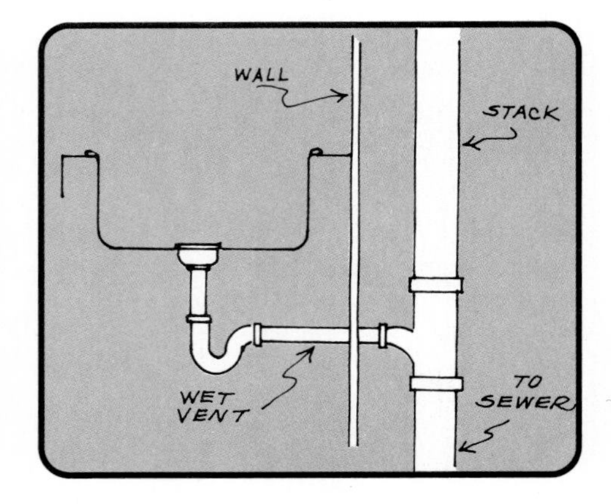

Wet venting lets a portion of the drain line serve also as a vent. Not all fixtures can be wet-vented (see above).

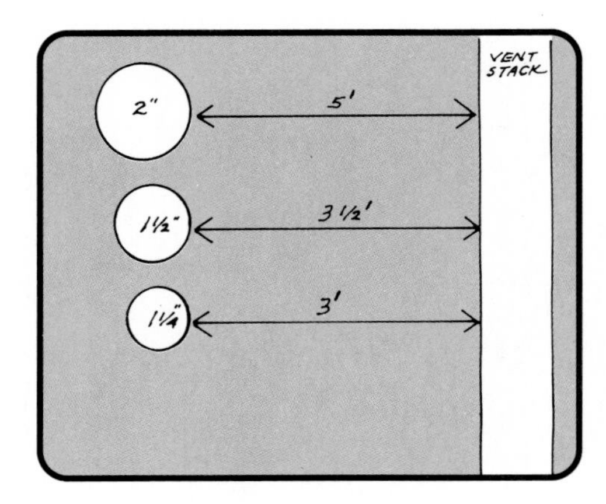

Maximum wet vent distances depend on the size of the fixture's drain. These are typical, but check your code.

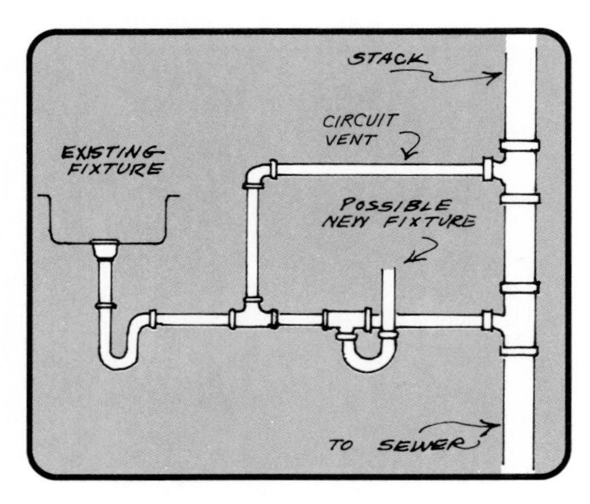

Existing fixtures may tie in via a *circuit vent*. If so, you can install a new fixture between the main and the circuit.

TAPPING INTO SUPPLY LINES

Saddle Tees

Looking for a way to run a new pipe off of an existing one without breaking a connection in the old run? Simply clamp on one of the saddle fittings shown at right, drill a hole, and hook up your new run.

Turn off the water first, of course, and open a faucet to drain the pipe. Next, strap on the saddle. Some come with drill guides so you can bore the hole without removing the fitting; with others, you mark the hole with a center punch, remove the saddle, and drill. File away any rough edges.

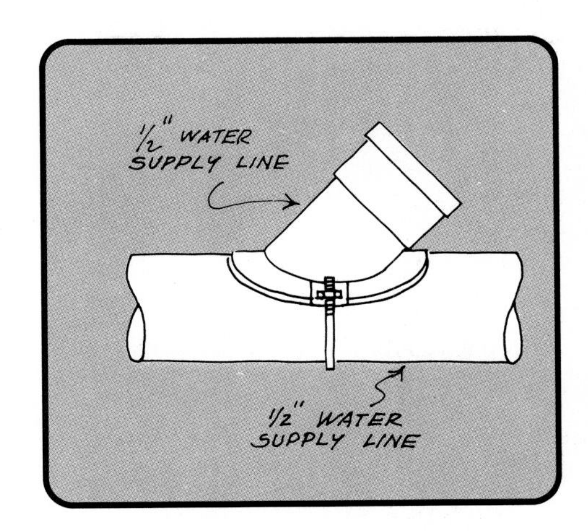

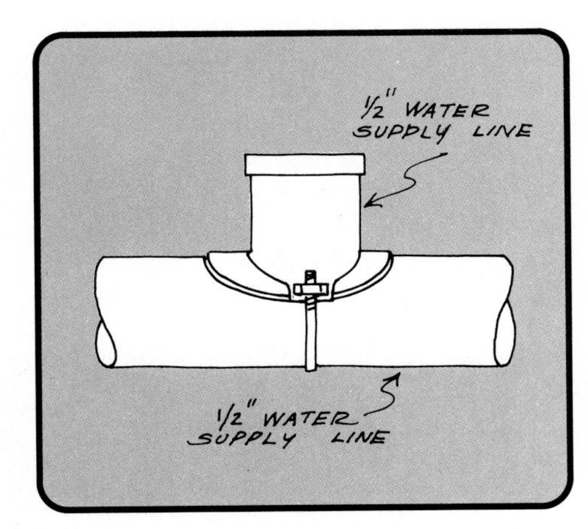

Slip Rings

Try to insert an ordinary rigid copper or plastic tee into an existing run and you may discover the line doesn't have enough flex to pop into the tee's socket. This means you'll either have to install a union, as explained below, or a slip ring, like the one shown here.

When you cut the line, remove a section big enough to accommodate the tee and a spacer that will help tie the tee into the existing line (see sketches at right). Fit the tee, spacer, and slip ring into the run, slide the ring into position, and solder or solvent-weld.

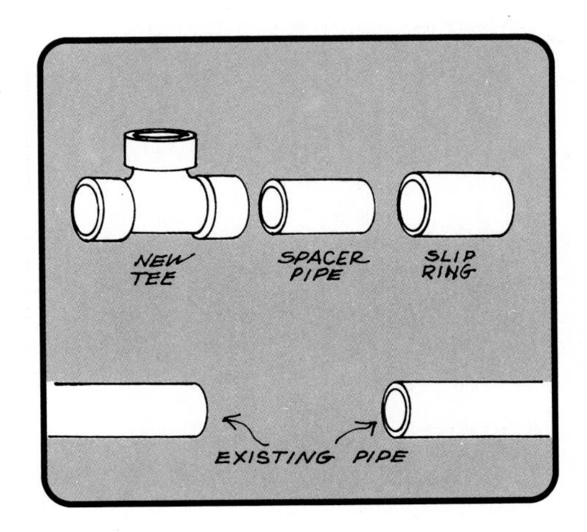

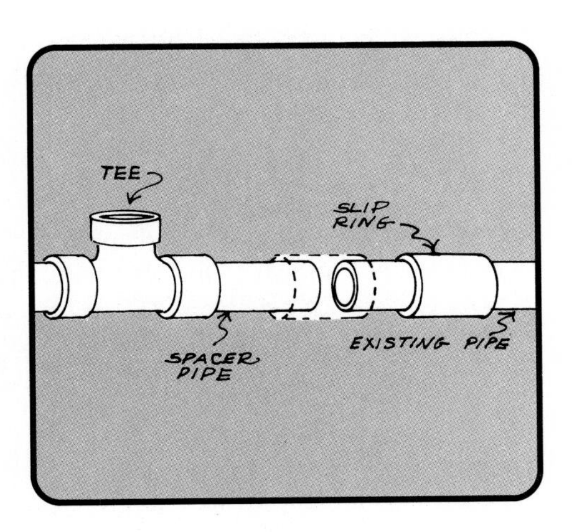

Unions

Unions are the keystones in threaded-pipe systems. These three-part devices compensate for the fact that all pipe threads in and out in the same direction.

If there's a union in the existing run, loosen its union *nut*, pull apart the fitting, and remove pipe on one side or the other. Now install a new tee and nipples, then reassemble as illustrated.

If there's no union in the existing run, simply cut in wherever you wish and add one. You may also need a second union at some point in the new run.

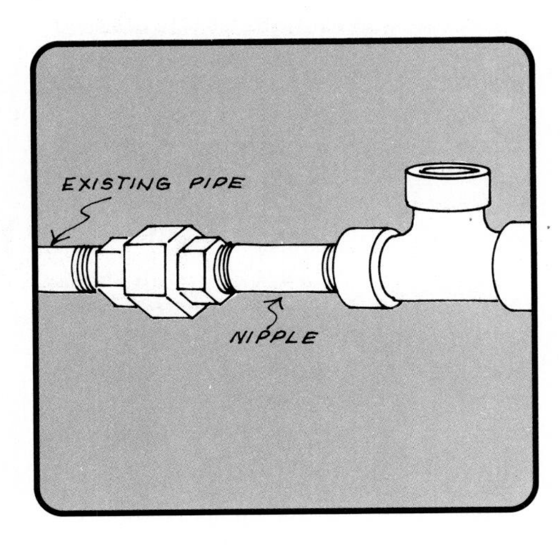

ROUGHING-IN

Plotting Dimensions

Once you know where and how you're going to tie into your home's existing plumbing system, it's time to pinpoint exactly where new lines will go. The best way to do this is to purchase your new fixture or fixtures then mark the "rough" dimensions on the floor and wall, as shown at right.

Don't let the word *rough* mislead you. Accuracy is essential—miss by an inch or so and you'll have to do a lot of work over again. These measurements are typical, but check your fixtures (or cut-outs that come with them) for specifics.

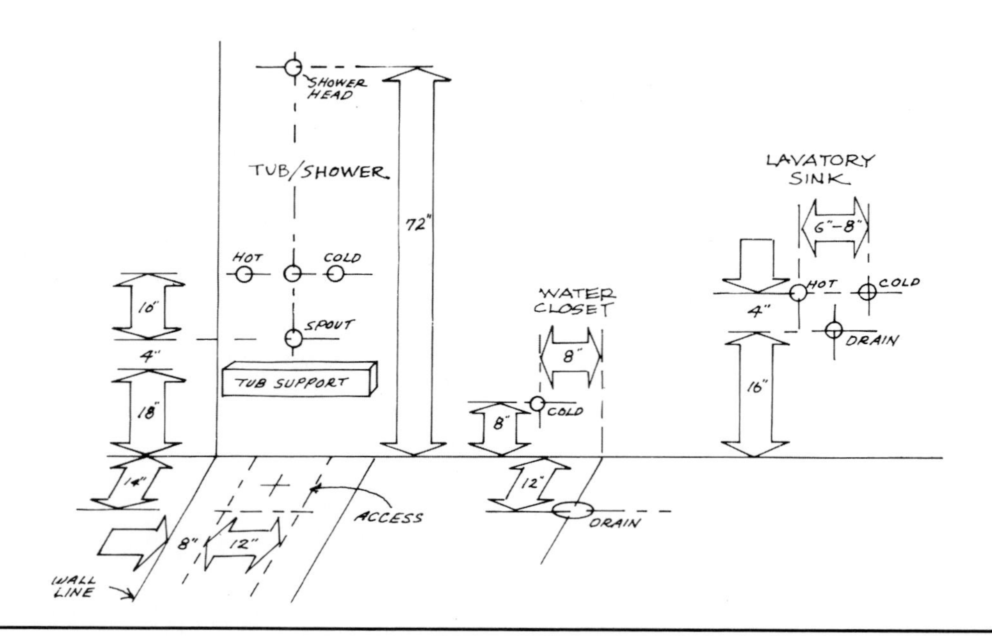

Supporting Exposed Pipes

Iron, steel, copper, and plastic pipes are tough customers, yet all of them must be supported to take the weight off threaded, soldered, or cemented connections. You can do this with any one of several types of pipe hangers, or with wood scraps or wire. Installation techniques are shown at right. Caution: If using metal pipe, make sure to use only hangers made of the same metal.

Unless otherwise specified by codes, space hangers every three feet. At points where pipes (especially cast iron pipes) make 90-degree turns, support the pipe with a piece of wood or use a strap hanger to support it.

Plastic and copper pipe need support more often than galvanized steel pipe. Although plastic and copper won't kink, they tend to belly-down when filled with water. This weight can quickly break connections and cause a serious leak. Note, too, that plastic needs more room for expansion and contraction, as explained on page 32.

If the pipe snakes along a masonry wall, the hangers sometimes can be installed with concrete nails. Drive the nails flush with the hangers, then stop hammering—extra"finishing" taps will loosen the nails. If concrete nails won't work, you'll have to use fiber plugs or lead anchors.

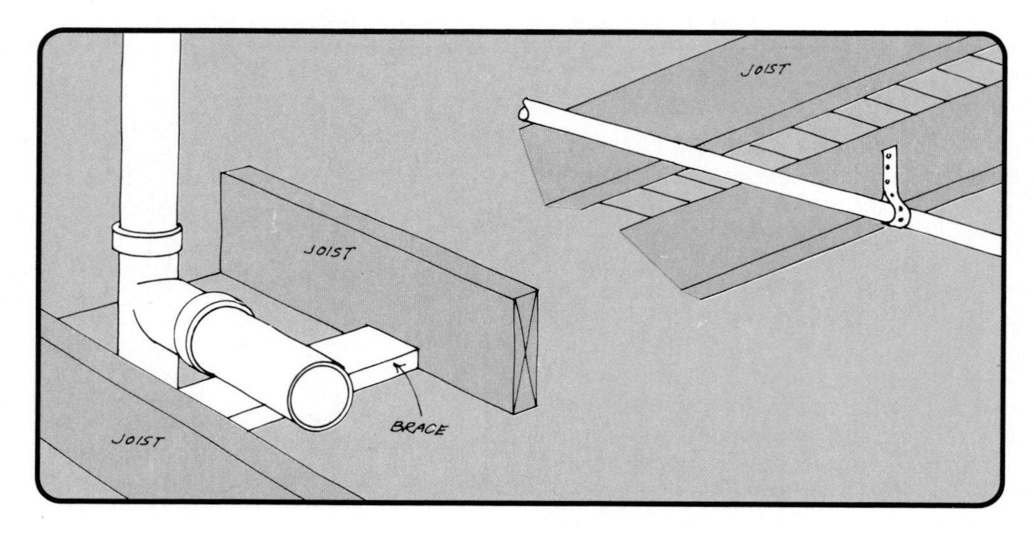

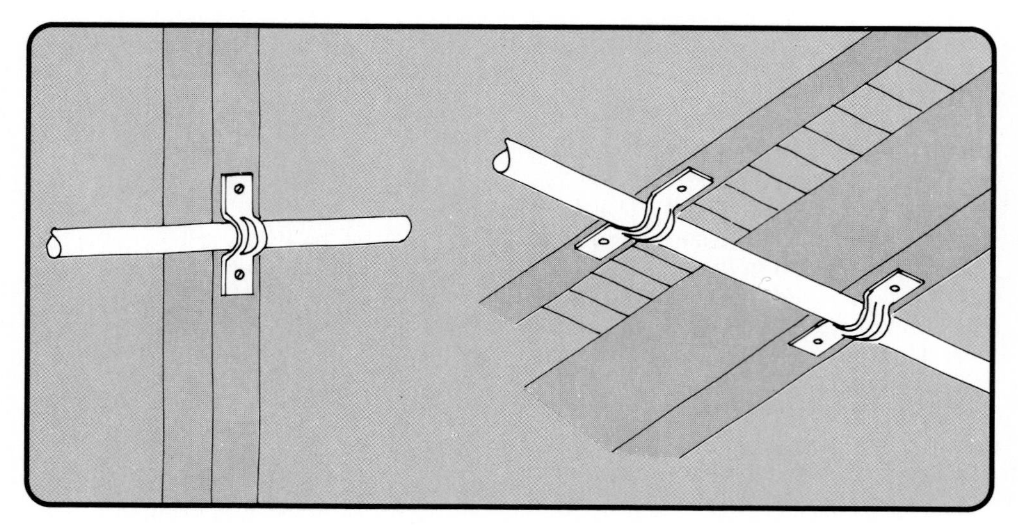

Notching Studs and Joists

Supporting pipes that pass through walls, under floors, or over ceilings is easy—you just cut notches in the framing members, or bore holes through them. Some plumbers prefer to notch; others bore. Generally, boring weakens a member less, but may create problems in assembling long runs.

Whichever method you choose, take care that you don't critically undermine a stud's or joist's load-bearing strength. Exactly how much you can safely cut away depends on where the notch or hole will be located.

If a notch will be in the upper half of a stud—usually four feet or more above the floor—you can safely cut away up to two-thirds of the stud's depth. In the lower half of a stud, though, don't notch more than one-third of its depth, unless you reinforce it as shown below. You can bore anywhere along a stud's length, provided you leave the clearance indicated.

With joists, never notch more than one-fourth of its depth—and keep the notch toward the ends, never near the center. Always reinforce joists, too.

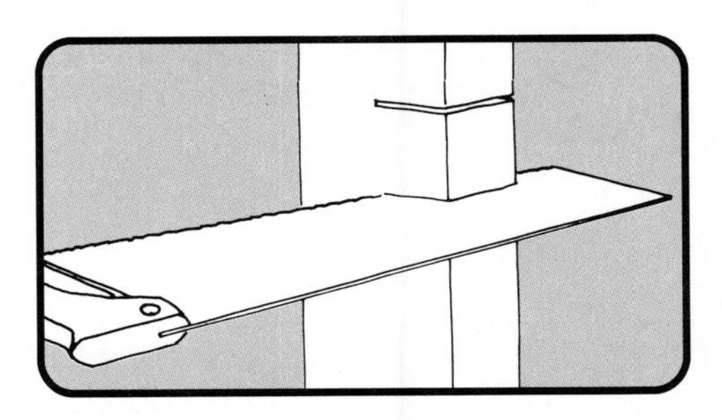

Make the cuts for notches with a handsaw or backsaw. Measure and lay out the cuts: don't guess at their width and depth.

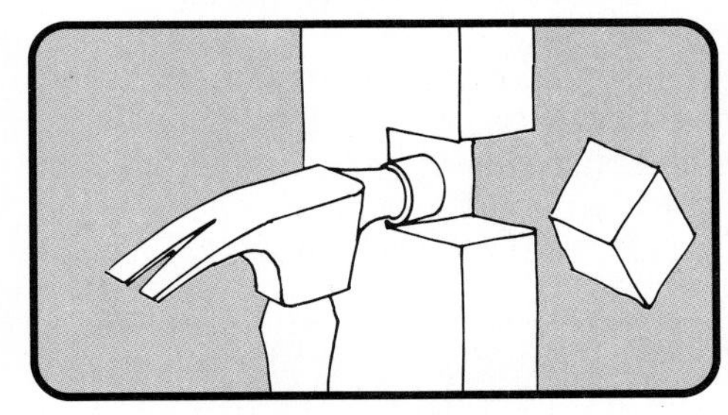

With a chisel, score a line on the framing member that connects the saw cuts, then knock out the notches with a hammer.

If the notch will be made next to a header or sill, use a hacksaw to make the cut. If you cut through a nail, renail the member.

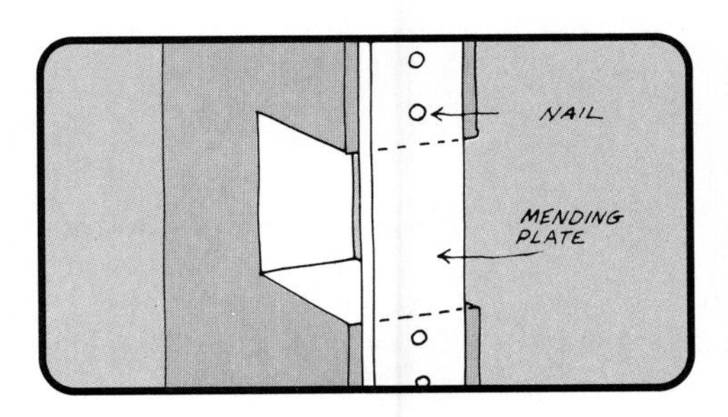

Reinforce deeply notched studs with a metal mending plate. You'll have to mortise it if the studs will be covered.

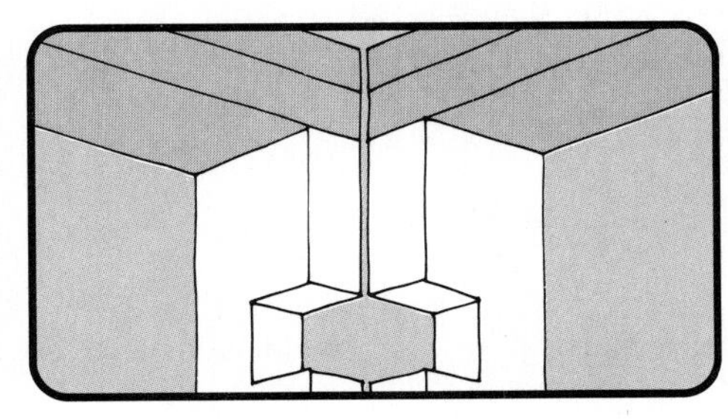

Notch a corner this way, taking half the notch from each framing member. A 90-degree pipe ell will fit snugly in the notch.

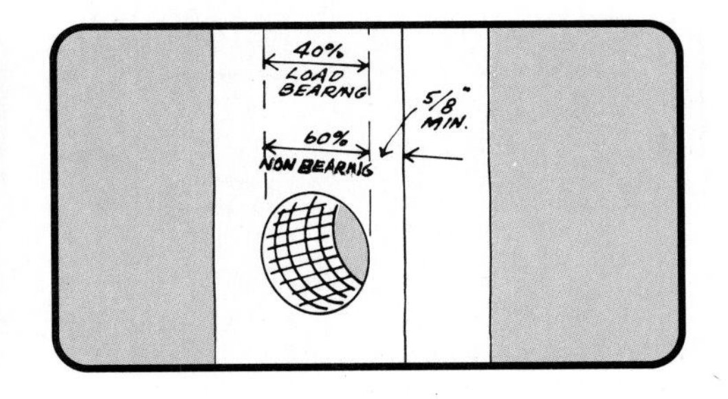

Leave at least ⅝ inch of wood around holes bored for pipes. The pipe should fit fairly snugly in the hole. Don't go oversize.

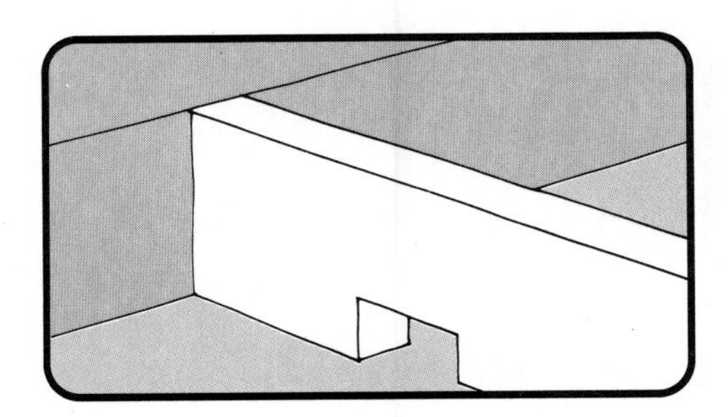

Notch joists near the ends, never near the center. Always cut the notch to fit the pipe. Big, sloppy cuts weaken the framing.

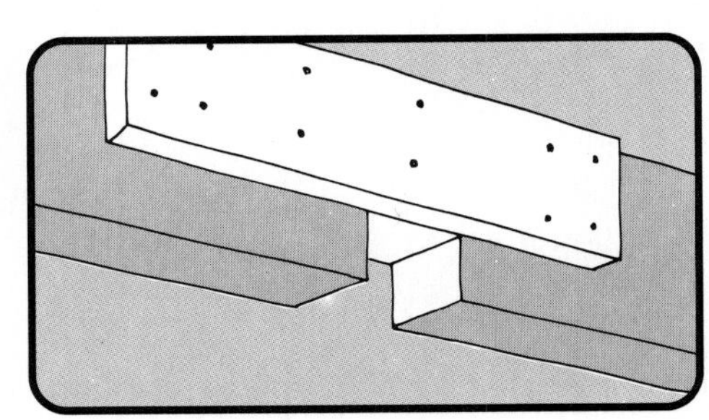

Reinforce joists with short lengths of wood nailed to both sides of the framing member. You can notch the patch, too.

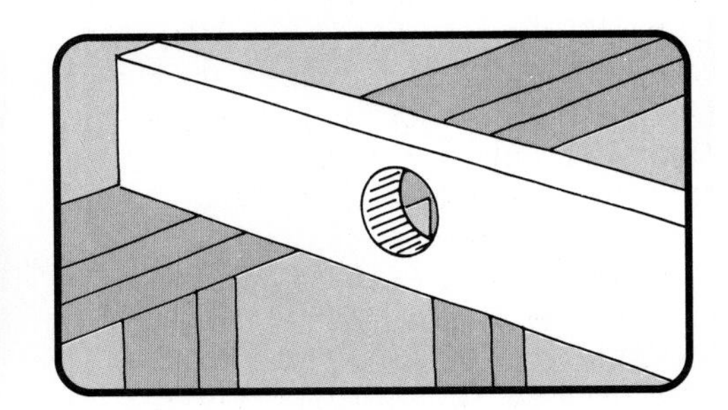

Keep the holes that you bore in joists near another support member to avoid undue stress on the joist.

INSTALLING SINKS AND LAVATORIES

Once you've roughed in new hot, cold, and drain lines for a new sink, about 75 percent or more of your work is completed. All that remains is to fasten a faucet assembly to the fixture (see page 42), mount the fixture itself, and fit it with a trap (see page 17).

Sinks and lavatories are generally made of steel (either enameled or stainless), porcelainized cast iron, plastic, or vitreous china. Deck-mounted types fit into a countertop or cabinet; wall-hung versions rest on brackets, and also may be supported by a pair of legs.

To install a deck-mounted sink, you first need to make a cutout for it. Most come with a pattern or basic dimensions you can use to make a template. Take care in laying out the cut, paying special attention to how the drain and supply lines will tie into the fixture.

Now cut the opening with a saber or keyhole saw; in a butcher-block counter, you may need to make pocket cuts with a circular saw, then finish up with a saw capable of making curved cuts.

Deck-mounted units may be either rim type or self-rimming, as shown below. With some models, you need to cut a recess around the opening. The manufacturer usually supplies full installation instructions. Follow these carefully to avoid difficulties.

Try to mount the faucet assembly before you set a sink or lavatory into place; with the bowl upside down you'll find it much easier to get at the faucet's locking nuts underneath.

Wall-mounting systems vary, but with most of them you fasten a bracket to the wall with toggle bolts, or first mortise in a hanging strip (see below). Make sure the brackets are absolutely level, then align slots or lugs in the lavatory's back and lower it into place.

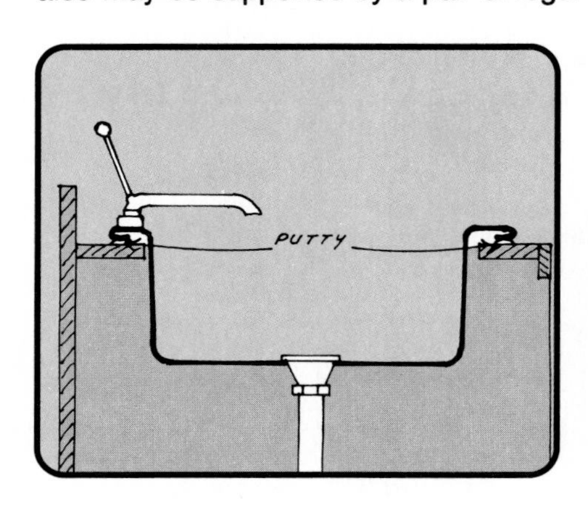

A self-rimming fixture has this profile. The counter or cabinet top may need to be recessed to accept the rim.

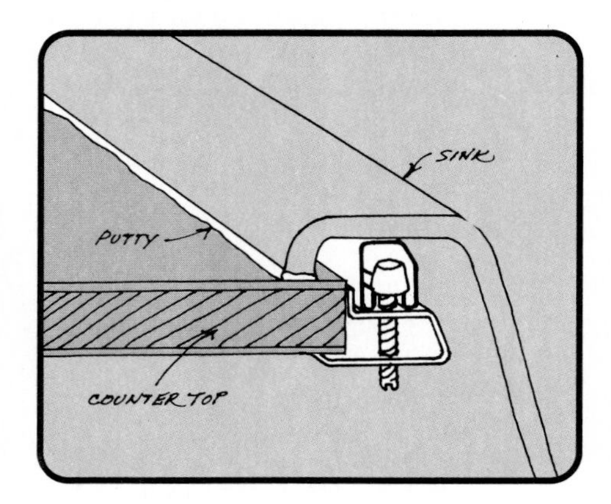

On lighter weight self-rimming sinks, screw-down clips similar to the one shown hold the sink tight against the countertop.

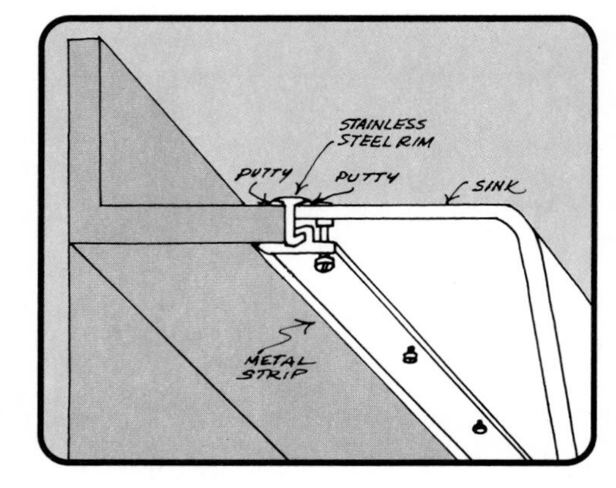

A rimmed fixture usually has a preformed metal strip in which the edge of the fixture is sandwiched. Secure with clips or screws.

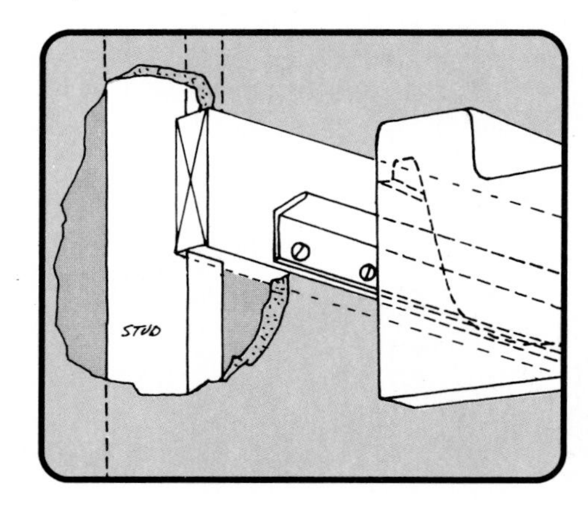

Recess a 1x8 into two wall studs for wall-mounting brackets. Position the face of the 1x8 flush with the finished wall.

Brackets are screwed to the 1x8 mounting piece. The style of the brackets depends on the fixture design. Brackets must be level.

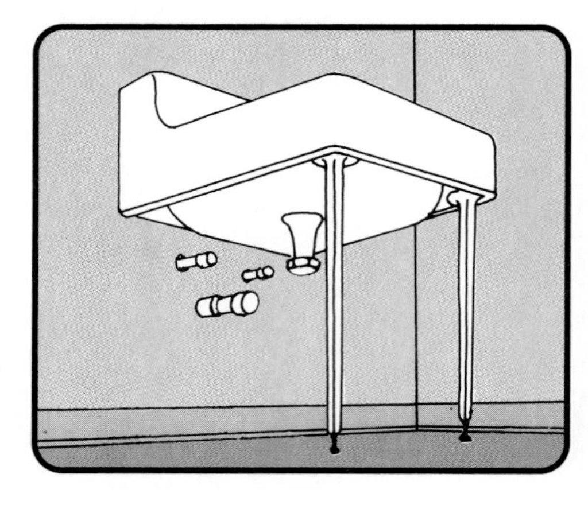

Attach the legs under the fixture apron. Unscrew them until they fully support the weight of the fixture.

INSTALLING TUBS AND SHOWERS

Of all plumbing fixtures, bathtubs and stall showers are the most "fixed." If you're constructing an all-new bath—or adding a basement shower—plan your framing around the installation and you can set in these bulky items without too much difficulty.

Start by selecting the unit you prefer. Standard tubs measure 4½, 5, and 5½ feet long—and you'll find lots of non-standard sizes, styles, and shapes to choose from.

You get a choice of materials, too. Baked-enamel steel tubs are relatively lightweight and inexpensive. They're prone to chipping, though, and can be noisy unless you insulate under and around the unit.

Cast-iron tubs are far more durable, somewhat more costly, and quite heavy. Plan to beef up floor joists under these. Though more expensive than either steel or cast-iron versions, fiber glass tubs usually include molded wall panels so you needn't worry about tiling or otherwise waterproofing around them.

For a shower, you can buy a standard *receptor base* and build your own enclosure, or purchase the entire unit base, walls, and a sliding or folding door—in knock-down kit form. The drawings below illustrate how each of these fixtures fits into new framing.

Thinking of replacing a deteriorated old tub with a new built-in? If so, prepare for a much bigger project. First comes the problem of getting the old tub out of a space that was probably built around it—major surgery that usually involves chopping into tile work and may entail removing a wall as well. Next you have to wrestle the new unit in, then adapt framing and plumbing to suit its dimensions.

Tying Into Framing and Plumbing

Check the anatomy drawing at right for the basics of a typical tub/shower installation. Drain connections call for dropping a trap below floor level—in space between joists or within a plumbing wall. If this isn't feasible, you'll have to elevate the tub a step or two. Position the faucets, spout, and shower head as indicated.

This tub, an enameled-steel type, has a flange around its edges that rests on 1x4 cleats and also is nailed to the studs. Recess another 1x4 into the studs to support the shower pipe.

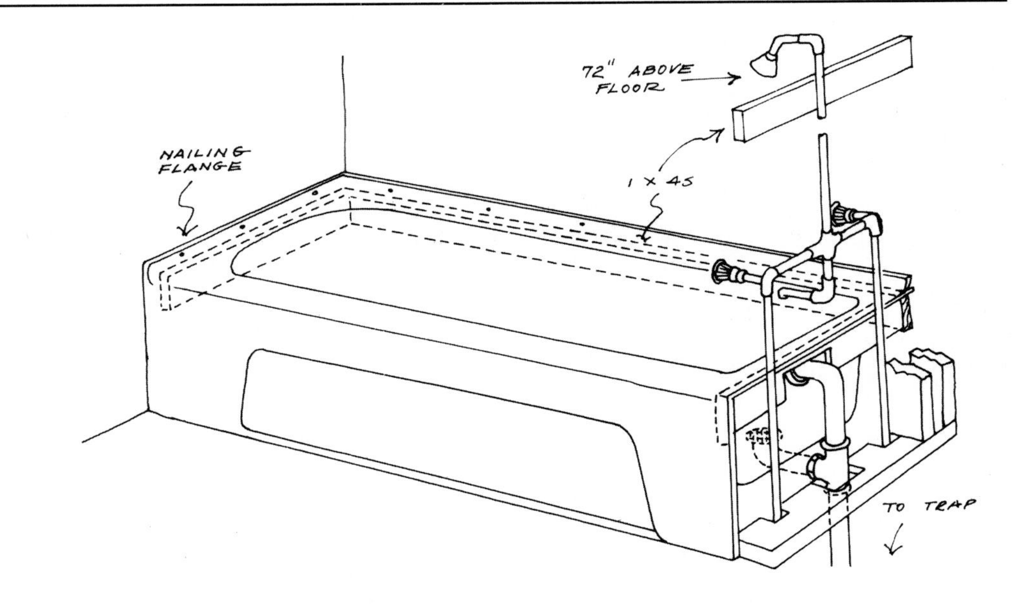

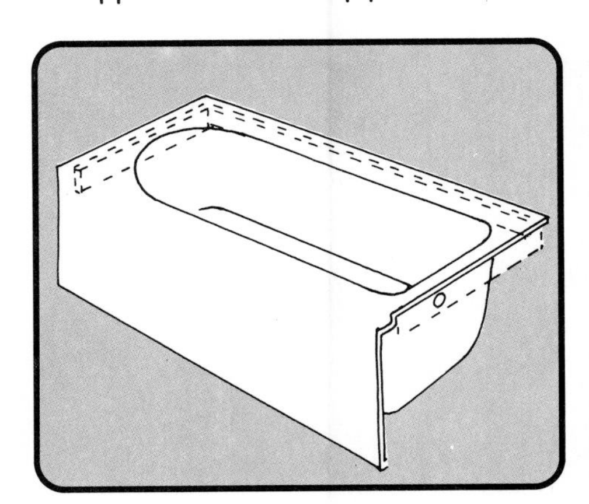

A cast-iron tub simply rests on 1x4 cleats. Its heavy weight keeps it stable, so you needn't nail it in place.

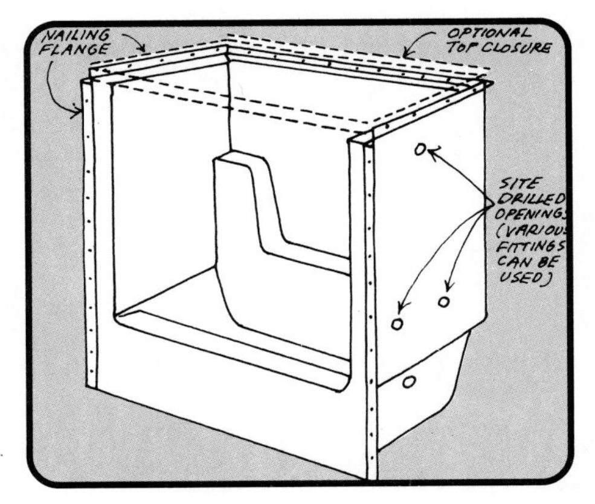

Molded fiber glass tub/shower units also have flanges for nailing to studs. Some dismantle to get into existing space.

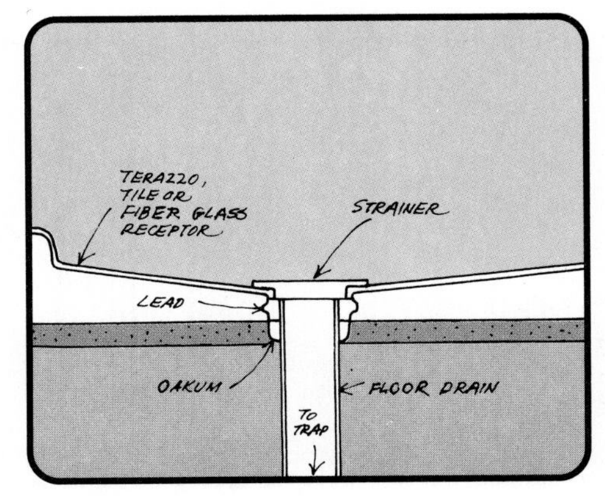

In a basement, you can position a shower base over a floor drain. Seal the connection with oakum and lead, as shown.

INSTALLING NEW FAUCETS

Selecting a new faucet is about like choosing a pair of shoes—first you pick out the style you like, then you make sure it fits properly.

If you're simply replacing an existing faucet, measure the distance between its pipe connections, as shown below. Or, better yet, disconnect the old unit and take it along to the store. The new faucet must fit the holes in your fixture exactly.

If, on the other hand, you'll be installing an entirely new fixture, choose the fixture first, then buy a faucet that's compatible. Don't worry about supply connections; flexible tubing connectors—often sold in kit form with shutoff valves—let you compensate for any differences here.

Disconnecting an old faucet can be tricky if the old connections are corroded or tough to get at. First, shut off the water and slip a bucket underneath to catch any water that may remain in the pipes. Now carefully fit a wrench to the connector nuts (you may need a basin wrench like the one shown below) and make sure the wrench has a good grip before you apply pressure—a slip could crack or dent the fixture.

If the connections won't budge, apply penetrating oil, wait for 20 minutes or so, and try again. As a last resort, heat the nuts with the flame from a propane torch, then turn them loose.

Faucets typically connect via a compression fitting that threads onto their inlets and is either threaded or soldered to the supply lines. If you have to solder, dismantle the faucet's working parts first (see pages 11–14) so they won't be damaged by the heat.

Deck-Mounted Units

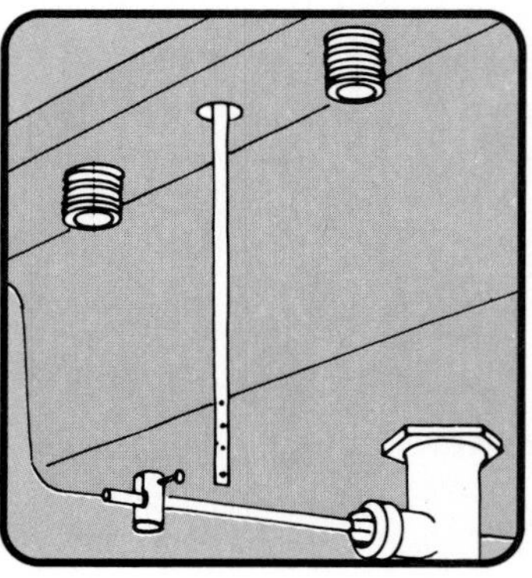

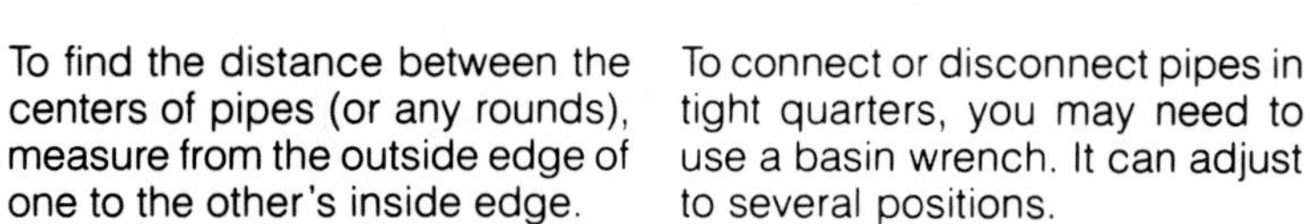

To find the distance between the centers of pipes (or any rounds), measure from the outside edge of one to the other's inside edge.

To connect or disconnect pipes in tight quarters, you may need to use a basin wrench. It can adjust to several positions.

A sink pop-up may be connected to the faucet. If so, remove the connection on the slide rod below the faucet unit.

Some copper tube connections have flared or compression fittings with nut connections. Take care not to bend the tubing.

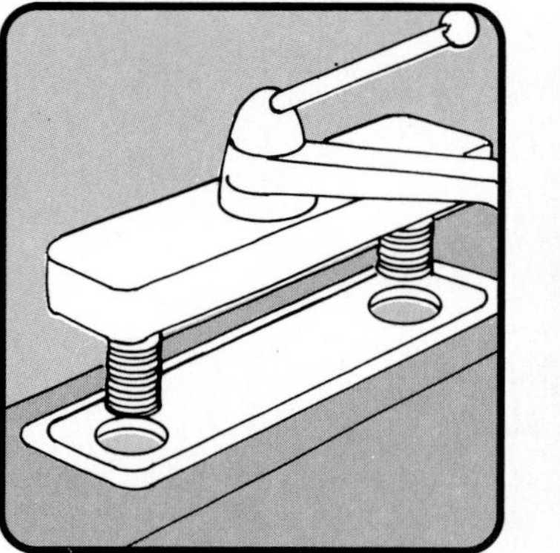

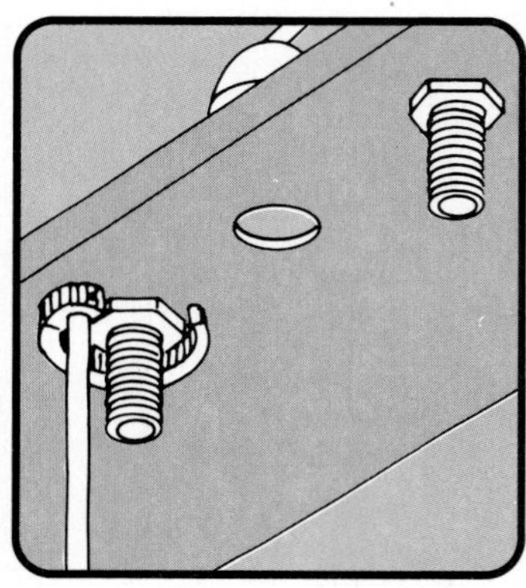

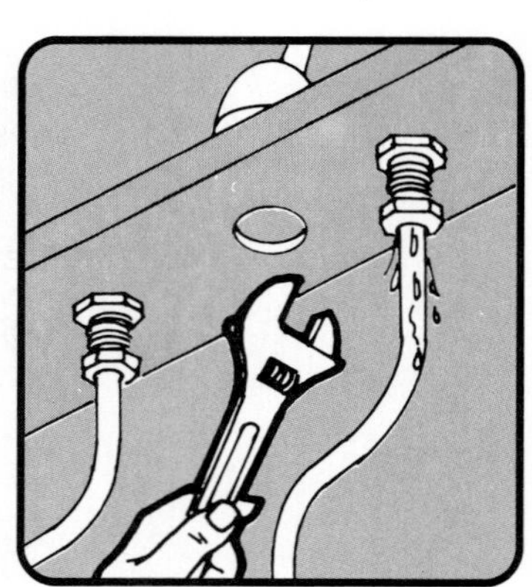

Insert the new faucet connections in the holes in the deck. Refer to the installation instructions for particulars.

Tighten the locking nuts below the deck, then make the water supply pipe connections. Use pipe dope or tape on threads.

If using copper tubing, bend it to fit the connections, but don't kink it. Then tighten the connections with a wrench.

Turn on the supply valves and test the pipes and faucet for any leaks. If you spot any, try tightening the connections.

Wall-Mounted Faucets

Though you gain access to wall-mounted faucets differently than deck-mounted faucets, the same installation techniques apply. Carefully measure the distance between the water supply lines, or take the old faucet to the store to match the fittings on a new faucet.

Since the fittings of wall-mounted faucets usually are plated with chrome or another finish, be careful when you turn them. A wrench or slip-joint pliers can leave marks. To prevent this, pad the jaws of tools with adhesive bandages.

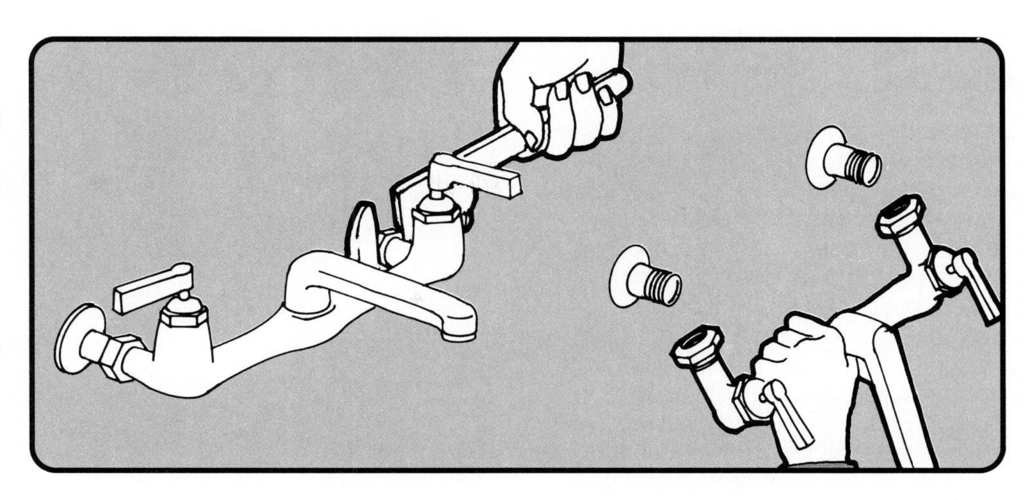

Tub/Shower Faucets

To replace or add faucets to shower and tub assemblies, you'll have to get in back of the finished wall to reach the connections. And unless the builder/plumber left an access panel for this purpose, getting there may mean poking a hole in the wall. Regardless of whether you cut into the wall from the front or the back, make sure the panel you cut allows you plenty of working room.

Once you gain access, close the shutoff valves controlling the water supply to the tub and disconnect the faucet. If you'll be removing soldered connections, protect the wall surfaces from heat and possible damage with a sheet of asbestos board.

Start by removing the faucet handles, escutcheons, and spout as shown. Don't mar plated parts.

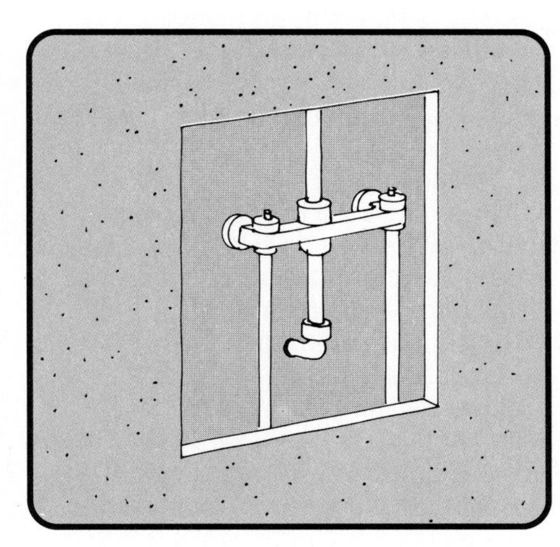

Cut an access hole in the wall with a keyhole saw. Cover it with a removable access panel.

Installing a Hand Shower

As you can see at right, hand showers can mount at either the shower head or the spout. Combination hand and stationary showers, though, connect to the shower pipe. These units have a bracket that holds a hand shower.

An existing shower head usually is connected to the water supply pipe with a threaded fitting. Simply loosen the fitting to remove the shower head. If you don't have the water supply pipe for a shower, you can replace your existing spout with a divertor type that has a connection for a hand shower.

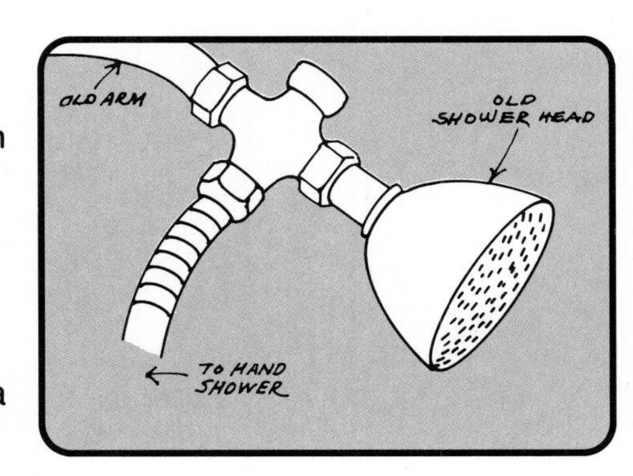

A standard or cross-tee fitting may be used for the addition of a hand shower. Make sure the tee fits all three connections.

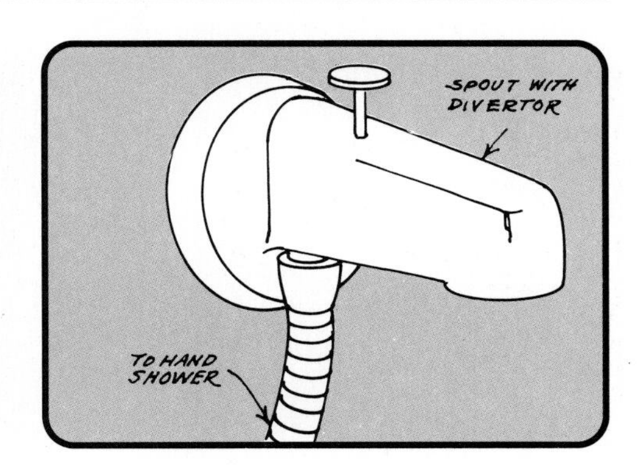

Divertor spouts also channel water to the hand shower. Make absolutely sure the new spout's threads match those of the pipe.

CHOOSING AND BUYING A TOILET

All of a toilet's mechanical action takes place in its *flush tank.* Since most of what can go wrong happens there, we show its components in an anatomy drawing on page 18, and devote the three pages that follow it to explaining how to make repairs.

When you're shopping for a new toilet, though, it also helps to know about the elements illustrated at right. Both the *tank* and *bowl* are molded of vitreous china, then fired in a high-temperature kiln. This produces a glaze that's impervious to just about anything but chipping, scratches, and cracks. Top-of-the-line versions combine the tank and bowl into a single piece, eliminating the *spud washer* and quieting the flushing action.

Flushing actions differ, too. All depend on water pressure to create a siphoning action—but some do this more efficiently and quietly than others. The drawings below compare the three basic types.

Regardless of design, all floor-standing toilets mount in the same way. A *closet flange* atop a *closet bend* accommodates *hold-down bolts.* These, plus the unit's considerable weight, maintain pressure on a wax or rubber *bowl seal* that prevents leaks. More about these on the opposite page.

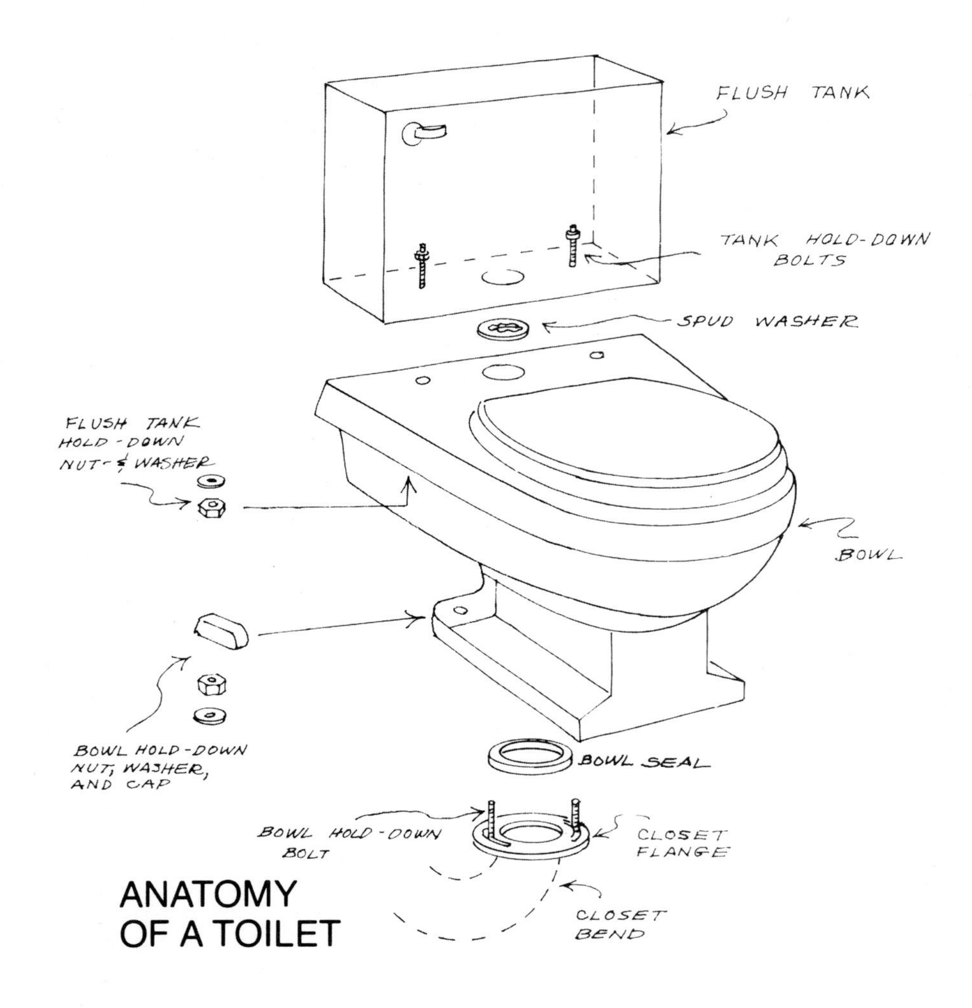

ANATOMY OF A TOILET

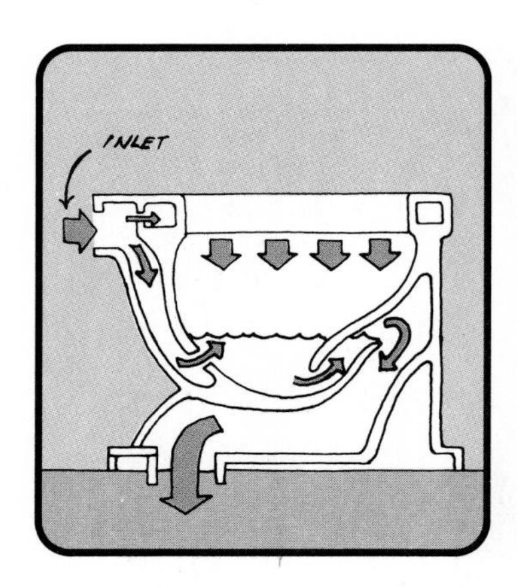

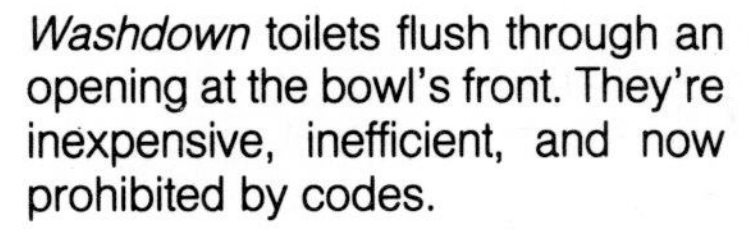
Washdown toilets flush through an opening at the bowl's front. They're inexpensive, inefficient, and now prohibited by codes.

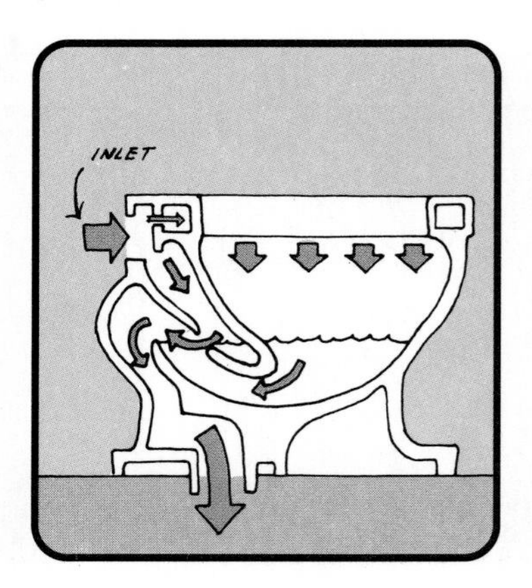

Better *reverse-trap* toilets flush through the rear for a quieter and much more efficient siphoning sort of action.

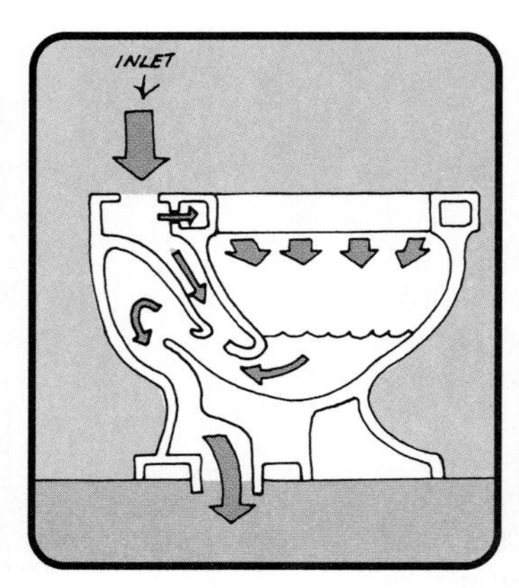

Siphon-jet types improve on the reverse-trap design with more water surface and bigger passages to reduce clogging.

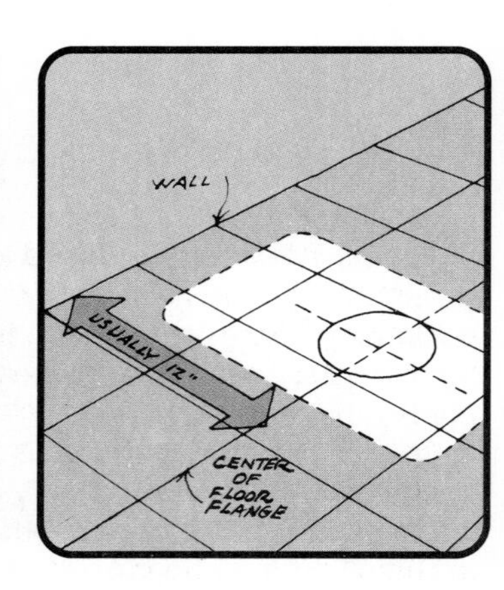

Before you buy a toilet, check this measurement. Most have a 12-inch "rough." A ten-inch version fits tight situations.

INSTALLING A TOILET

Whether you're replacing an existing fixture or mounting a new unit on a closet bend that's already in place, setting and hooking up a toilet is a surprisingly simple operation.

Besides the toilet itself and a few ordinary plumbing tools, you'll need some plumber's putty, a bowl seal, and—for a new installation—a closet flange.

With an existing fixture, first shut off and disconnect the water supply, flush the tank, then swab out water that remains in the tank and bowl with an old towel or sponge.

Next, check to determine if the tank is connected to the wall. Some older units attach via screws through the rear and have an elbow connection to the bowl. If that's the case with yours, disconnect these and remove the tank. If the tank sits atop the bowl, the toilet can be removed in one piece, though you may need help to lift it.

Now pry off the caps that cover the bowl's hold-down bolts. Remove the nuts—or cut the bolts with a hacksaw—and just lift the bowl off the flange. Get it out of the way so you'll have room to maneuver the new unit.

Before you install the new bowl, temporarily set it on the flange and check for level—front to back as well as side to side. Shim, if necessary, with rustproof metal washers.

Now lift the bowl off again, turn it upside-down (on padding to protect the rim), and fit the seal around the bowl's outlet. Run a bead of putty around the outer rim of the bowl's base, too, so dirt and water won't get underneath.

Set the bowl in place as illustrated below, check for level once more, and install nuts on the hold-down bolts.

Finally, fit a spud washer and the tank over the bowl's inlet opening, secure with bolts, and hook up the water supply. If you don't have one, now's the time to install a shutoff, too, as shown at the bottom of the page.

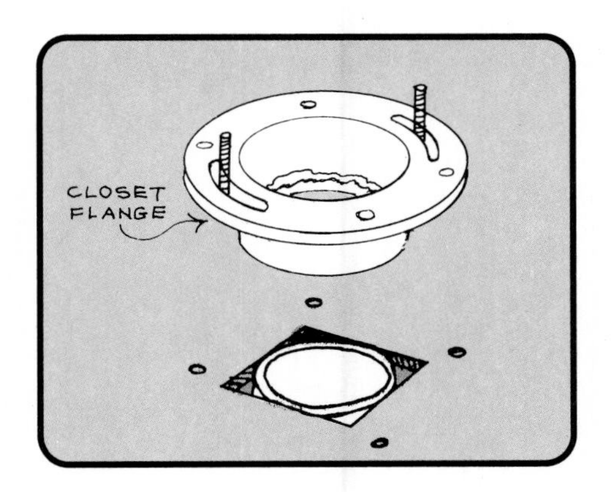

Apply cement to both the closet bend and the flange as shown, press the flange into place, and screw it to the floor.

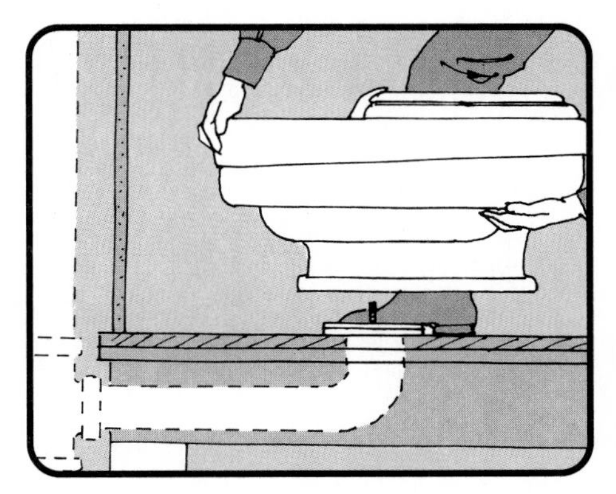

Set the bowl in place with a slight twisting motion, but don't rock or lift it again—you could break the seal.

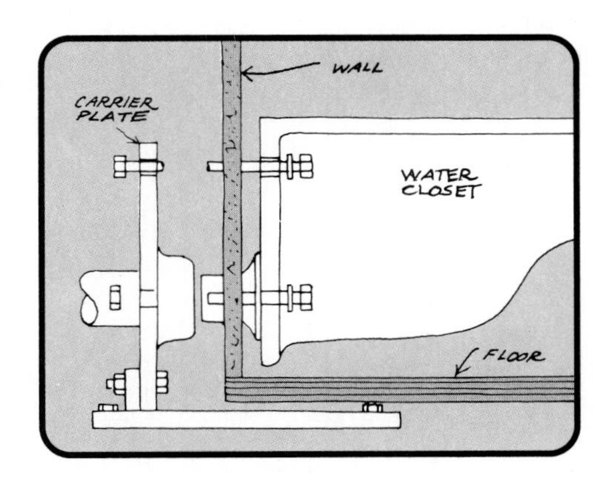

Wall-mounted toilets attach to a carrier plate behind the wall. These have a similar flange-and-gasket arrangement.

INSTALLING FIXTURE SHUTOFFS

Tired of turning off your entire water system every time you need to replace a faucet washer or work on a flushing mechanism? Fixture shutoffs—often sold in easy-to-hook-up kit form—spare hassles and trips to the basement.

Before you set out to buy the parts, measure the supply pipe or pipes you'll be attaching to and note whether they're threaded steel, copper, or plastic. Each requires different fittings or adaptors. For installation, refer to the appropriate pipe-fitting techniques on pages 26–33.

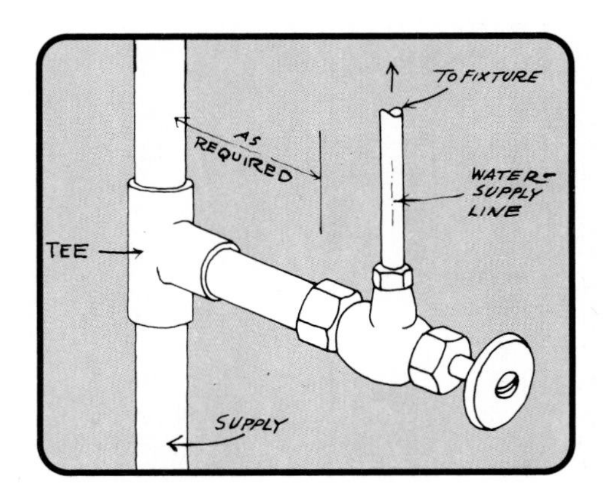

Angle stops suit cases where pipes come out of the wall. Flexible tubing with compression fittings connects to the fixture.

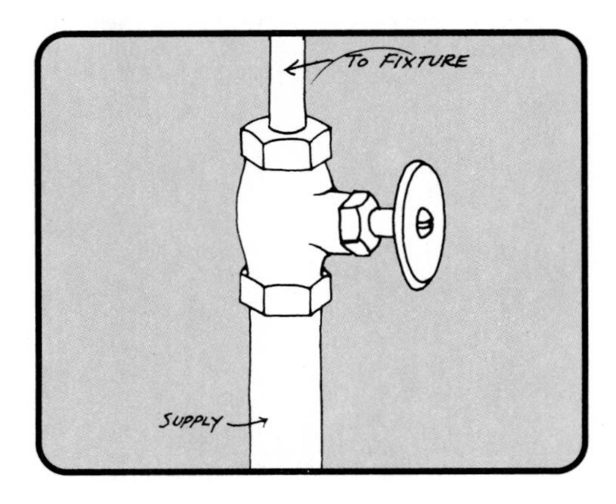

If pipes exit from the floor, choose an *in-line* stop. Stop outlets are usually smaller in diameter than house plumbing.

INSTALLING A WATER HEATER

Most water heaters serve dutifully for many years, but sooner or later all succumb to rust and corrosion and need replacement. When you find yourself faced with this situation, consider whether or not you really need a pro to install a new one.

There's nothing especially difficult about hooking up a water heater. Even if you want to increase—or decrease—your water heating capacity, you can usually find a replacement with almost the same overall dimensions as the old equipment. This spares you the trouble of retailoring plumbing lines.

Before you rush out to get a new heater, though, make sure it's really the tank that's leaking—not the overflow or a poor connection. (To learn about troubleshooting water heaters, see page 22.)

If the old heater wasn't keeping up with your family's needs, consider upgrading to a unit that holds the same amount of water but has a faster "recovery rate." Higher recovery rates deliver more gallons of hot water during peak periods, without wasting energy during idle times.

When your new unit arrives, arrange to have a helper on hand. Traversing stairs with a tank that weighs 125 to 200 pounds can be tricky. In some cases, you might be better off to remove a basement window, set up a ramp with planks, then slide the new unit in and the old one out.

After you uncrate your purchase, set it up next to the old heater, study the installation instructions that come with it, and determine whether you need any new plumbing or flue fittings. Some economy-minded plumbers don't bother to install unions. If that's the case at your house, you'll have to cut the lines to remove the water heater. To save trouble in case anything goes wrong in the future, invest in a couple of unions—and a shutoff valve, too, if you don't already have one.

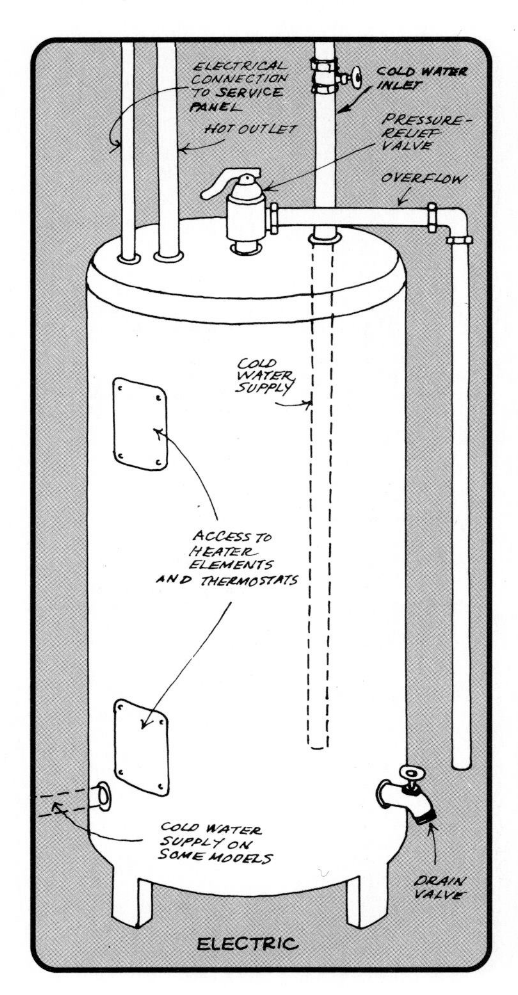

An electric water heater has four hookup points: hot and cold water supply lines, electric power, and a pressure-relief valve.

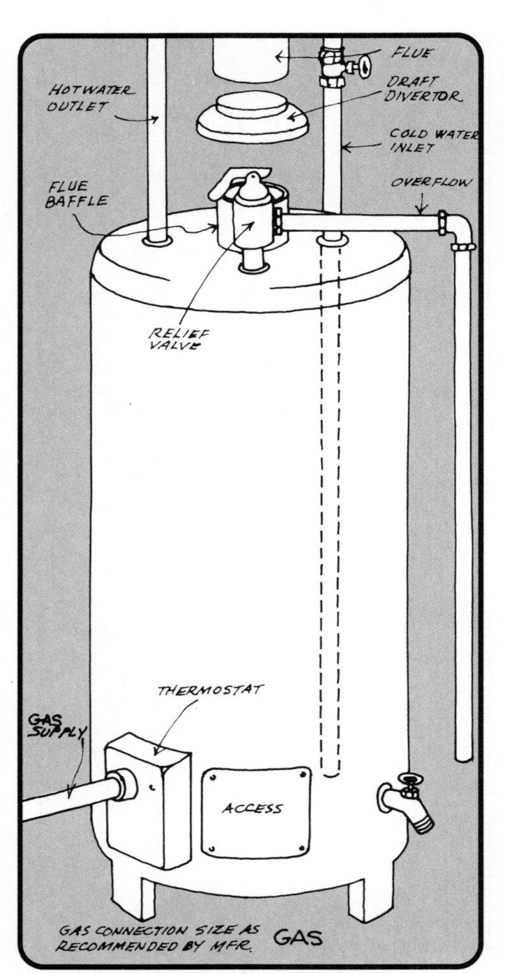

A gas water heater, in addition to hot and cold water hookups and a relief valve, also has a gas line and a flue stack.

Turn off the water at the shutoff valve near the heater or at the water main. You *must* do this before starting any other work.

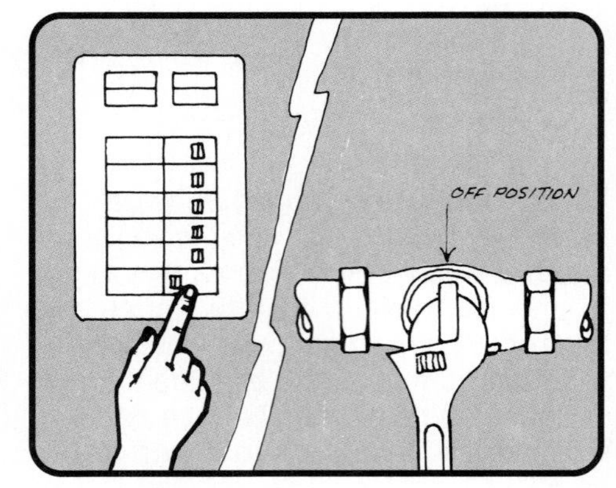

Shut off the power at the service panel, or halt the supply of gas by closing the shutoff valve leading to the water heater.

After the water to the tank has been turned off, drain the tank. This can take several hours, so connect a hose to the drain valve.

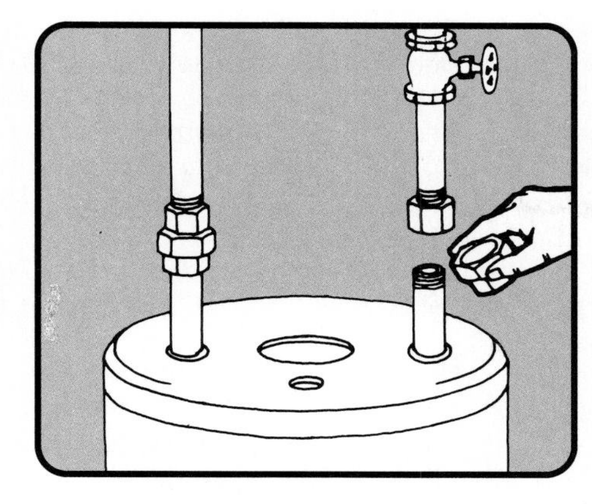

Disconnect the water and gas lines. See pages 29 and 31 if you have to cut the pipes. Slide the old unit out of the way.

Don't make any connections until the new heater is level. You can level it with cedar-shingle shims or pieces of rot-resistant wood.

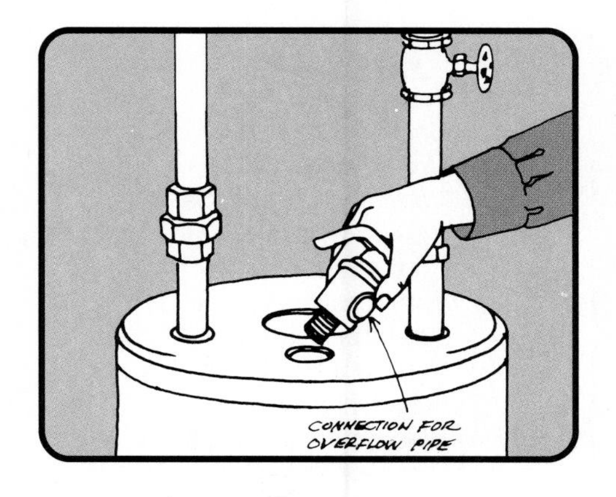

Connect the pressure-relief valve and reconnect the power source to the heater. You may be able to use the old connection.

Install the draft divertor and position it over the flue baffle. Then connect the flue pipe. The parts usually are in the new kit.

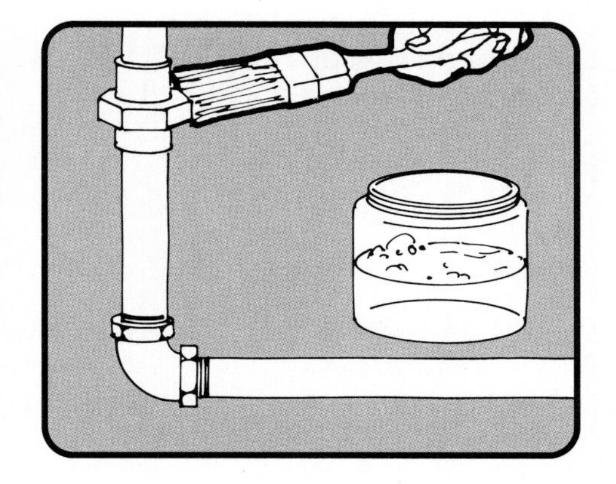

On new connections, check for gas leaks with a solution of soap and water. Use an old brush to swab the joint; look for bubbles.

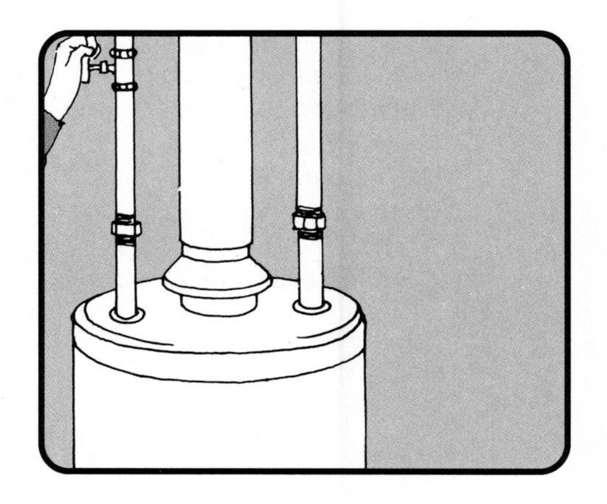

Turn on the water and fill the tank. Then open a hot water faucet in the kitchen. When the water flows, fire the unit.

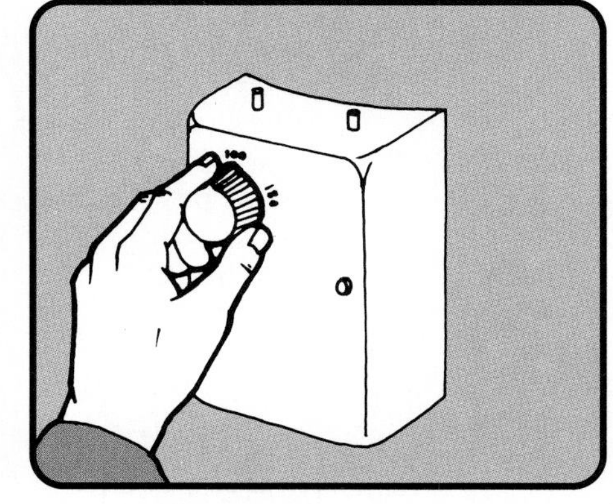

Set the thermostat at about 140 degrees F. after the power is on. Lowering the setting a few degrees will save some energy.

Drain the new hot water heater every two months the first year; every six months afterward. Draw off about two gallons of water.

INDEX

A–I

J–P

R–Z